Alice Merritt Davidson

California Plants in their Homes

A botanical Reader for Children

Alice Merritt Davidson

California Plants in their Homes
A botanical Reader for Children

ISBN/EAN: 9783337176693

Printed in Europe, USA, Canada, Australia, Japan

Cover: Foto ©ninafisch / pixelio.de

More available books at **www.hansebooks.com**

YUCCA WHIPPLEI.

CALIFORNIA PLANTS
IN THEIR HOMES

A BOTANICAL READER
FOR CHILDREN

BY

ALICE MERRITT DAVIDSON

Formerly Teacher of Botany in the State Normal School,
Los Angeles, California.

ILLUSTRATED
BY
ALICE C. COOPER
AND
MARY E. LEWIS

1898
B. R. BAUMGARDT & CO
LOS ANGELES, CAL.

COPYRIGHT, 1898
BY
ALICE MERRITT DAVIDSON
74444

CONTENTS.

	PAGE.
LIST OF ILLUSTRATIONS	10
PREFACE	13
CHAPTER I	17
Some Plants that lead Easy Lives.	
CHAPTER II	25
How Some Plants begin Life: Seedlings.	
CHAPTER III	39
Plants that know how to meet Hard Times: Autumn Plants.	
CHAPTER IV	51
Some Plants that do not make their own Living: Fungi.	
CHAPTER V	61
After the Rains: Winter Plants.	
CHAPTER VI	72
Ferns and their Relatives.	
CHAPTER VII	85
Some Early Flowers.	
CHAPTER VIII	100
The Awakening of the Trees.	
CHAPTER IX	110
Some Spring Flowers.	
CHAPTER X	126
Plants with Mechanical Genius: Lupine, Alfalfa and Filaree.	
CHAPTER XI	138
Plants of High Rank: Bilabiate Flowers.	
CHAPTER XII	151
Social Flowers: Compositæ.	
CHAPTER XIII	160
Plant Families: Part I. Endogens or Monocotyledons.	
CHAPTER XIV	172
Plant Families: Part II. Exogens or Dicotyledons.	
CHAPTER XV	183
Some Summer Flowers.	
CHAPTER XVI	200
Weeds.	
Errata	212
Books of Reference	213
Pronunciation of Botanical Names	215

ILLUSTRATIONS.

FIG.		PAGE
1.	Water-net—*Hydrodictyon*..	18
2.	A Brown Alga, or Kelp—*Macrocystis*..................................	20
3.	A Red Alga—*Plocamium coccineum*......................................	23
4.	Some Green Algæ..	23
5.	Castor-oil Plant—*Ricinus*, Germination of Seed...................	26
6.	Germination of Pine Seed..	28
7.	Germination of Some Common Dicotyledonous Seeds..........	30
8.	Germination of Acorn...	31
9.	Germination of Some Monocotyledenous Seeds....................	35
10.	Castor-oil Plant, Cellular Structure.....................................	36
11.	Wild Broom—*Lotus glaber*...	45
12.	*Eriogonum elongatum*...	45
13.	A Summer Composite—*Malacothrix tenuifolia*....................	46
14.	Turkey-weed—*Eremocarpus setigera*.....................................	47
15.	Plants of the Dry Season..	40
16.	Plant Hairs under the Microscope.......................................	50
17.	Mould..	53
18.	Lichens...	55
19.	Some Common Fungi...	58
20.	Early Seedlings...	62
21.	Seedlings spread out to the Sun..	63
22.	Early Growth from Underground Storehouses.....................	65
23.	Chilicothe—*Micrampelis macrocarpa*.....................................	67
24.	Wild Currant—*Ribes glutinosum*..	69
25.	Wild Gooseberry—*Ribes amarum*...	70
26.	*Polypodium Californicum*..	73
27.	Some Early Fern Fronds...	76
28.	Fragments of Spore-bearing Fronds.....................................	78
29.	Scouring Rush, or Horsetail—*Equisetum*..............................	82
30.	Young Ferns and Liverworts...	83
31.	Peony—*Pæonia Californica*...	86
32.	Buttercup—*Ranunculus Californicus*...................................	88

ILLUSTRATIONS.

33. Cluster Lily—*Brodiæa* **Capitata**... 91
34. Violet—*Viola pedunculata*... 94
35. Shooting Star—*Dodecatheon* **Clevelandi**............................ 96
36. Calla... 98
37. Willow.. 101
38. Young Shoots of Trees.. 103
39. Peach Blossoms.. 105
40. Pine and Cypress.. 106
41. Poppy and Cream-cup... 111
42. Primrose—*Œnothera bistorta*.. 113
43. Gilias.. 115
44. The Climbing Nemophila—*Nemophila aurita*..................... 117
45. Some Members of the Blue-Eyes Family............................. 118
46. White Forget-me-not—*Plagiobothrys nothofulvus*............... 120
47. Nightshade—*Solanum Douglasii*... 122
48. Mariposas... 124
49. Lupine—*Lupinus sparsiflorus*.. 127
50. Alfalfa and Bur-Clover.. 130
51. Filaree—*Erodium cicutarium*... 134
52. Monkey-Flower—*Mimulus glutinosus*.................................. 139
53. Innocence—*Collinsia bicolor*... 141
54. Owl's Clover and Painted **Cup**... 143
55. Violet Pentstemon—*Pentstemon heterophyllus*................... 145
56. Little Chia—*Salvia Columbariæ*... 147
57. Sages—*Audibertia stachyoides* and *A. polystachya*............ 149
58. Sunflower—*Helianthus annuus*... 152
59. Thistle—*Cnicus occidentalis*... 156
60. An Umbellifera—*Peucedanum utriculatum*......................... 158
61. Blue-Eyed Grass—*Sisyrinchium bellum*............................... 165
62. Yucca—*Yucca Whipplei*. (See also Frontispiece)................167
63. Wild Oats—*Avena fatua*... 169
64. Yerba Mansa—*Anemopsis Californica*............................... 173
65. Larkspur—*Delphinium Parryi*... 185
66. Milkweed—*Asclepias eriocarpa*.. 188
67. Cactus—*Opuntia Lindheimeri*, var. *occidentalis*................ 190
68. Dodder on *Eriogonum fasciculatum*.................................... 192
69. *Godetia Bottæ* and *Clarkia elegans*.................................... 195
70. Indian Pink—*Silene laciniata*.. 195
71. Climbing Pentstemon—*Pentstemon cordifolius*................... 196
72. Pasture Weeds.. 206
73. Wayside Weeds.. 209

For alphabetical list, see index.

PREFACE.

The aim of this book is to foster children's love for nature and for out of door life by stimulating their interest in living plants, and by leading them to study the habits of their plant neighbors. Out of door life is possible in California all the year round, but to avail one's self fully of this great panacea for mental and physical ills, one must have capacity for enjoying it. There is no stronger incentive to this sort of recreation than genuine enthusiasm for some branch of nature study. No one who has gone into the fields with children, doubts their readiness to acquire this resource, but there must be guidance that leads to keener observation, to a growing acquaintance with nature and a deepening interest in her ways. A book can hardly meet the full requirements of a guide, but supplemented by the direction and sympathy of parent or teacher, it may be of much service.

Of course these "plants in their homes," should be under the actual observation of the children. The only plants treated at all fully are such as the author has found easily accessible in Southern California, and most of them are common throughout the state. That the study of some of them necessitates special excursions to cañon or hillside, is no objection. The greater number, however, are the plants we meet without leaving beaten paths; the filaree and bur-clover, for instance, serve often for illustration, and

PREFACE

our common weeds have a chapter to themselves. The aim is to awaken interest in common plants, and to invest them with new meaning. Pond-scums, mould and toad-stools are included among these familiar plants, and sea-mosses, lichens and ferns are believed to be as attractive as flowering plants. Much concerning the lives and habits of these lower plants can be learned without a microscope, but the results of microscopic study of these and of higher plants are not ignored in the Reader.

The book attempts to introduce children to a wide circle of plants. Entire, living plants are considered, and leading facts concerning their physiology and their relations to their environment are pointed out. A detailed study of the structure of plant fragments is not suggested. The value of rigorous laboratory training to mature minds is undeniable, but experience seems to prove that the methods of a college laboratory are not suited to children. In plant study children can be led to see and seek and think for themselves, but their natural interest is in salient features not in minute details, and their curiosity is concerning uses, not structure.

In California, parents and teachers who would gladly undertake rational plant study with their children, are confronted by our peculiar climatic conditions, which render the courses of study and the plant literature of other regions ill adapted to our needs. It is in response to many expressions of need for some literature of our own in this line, that this book has been undertaken. It records observations made during ten years of much out of door life in Southern California. They are offered, not for the sake of any definite results obtained, but for the purpose of stimulating like observations and for comparison. Much auxiliary matter, directions, definitions, physiological facts and theories, and the like, must accompany observation work to render it intelligible. In school work this matter must

PREFACE

be dictated or written by the teacher. The Reader is designed to save this labor, that is, to supplement observation work, but by no means to supplant it.

The Supplement to the Reader has been written with the needs of teachers in mind. It contains many additional details; the plants are described more fully and their botanical names are given. Such facts of structural and physiological botany as the writer has found specially useful during considerable experience in teaching, have been inserted. In an appendix, suggestions for the use of the Reader in the school-room are offered, and a course of plant study for California school children is outlined. A short list of available botanical works is also given.

In the preparation of the book, the usual botanical authorities have been consulted. Kerner's "Natural History of Plants" has suggested and inspired investigation along various lines. Direct assistance in every way has been given by the writer's husband, Dr. Anstruther Davidson. Many thanks are due to Miss Harriet E. Dunn for aid in revising the manuscript and to Mrs. J. Crossley Neilson and Prof. Everett Shepardson for assistance in proof reading. The drawings, nearly all of them from nature, have been contributed by former pupils in the Normal School at Los Angeles, most of them being the work of Miss Cooper and Miss Lewis. Miss Ada M. Laughlin, teacher of drawing in the same school, has been most helpful.

FOOT NOTE.—It has not generally been practicable to reproduce the drawings natural size, or to an exact scale. Most of the flowering plants are reduced about one-fourth. The number of times magnified is indicated by the sign, x.

CHAPTER I.

SOME PLANTS THAT LEAD EASY LIVES.

There is a story told of a very lazy boy who lived in a beautiful sunny garden. The garden was full of trees that bore delicious fruit, but to gather this fruit was quite too much trouble for the lazy boy; so he spent his days lying under the trees where, when he was hungry, he had only to open his mouth and the fruit dropped into it. Now there are plants the world over that seem to live about as easily as the lazy boy; I mean plants that live in the water, for their food consists only of water and what is dissolved in it; so they lie in a bath containing their food and have only to absorb what they need.

Nearly all California children have noticed in reservoirs, water ditches or ponds, a green scum floating along the edges or on the top. Perhaps you have not looked at it very closely, nor thought of it as made up of little plants. Take some of it now, put it in water in a white dish, and examine it carefully. Perhaps all of it will be soft and slimy, but you are likely to find some that clearly consists of threads or nets.

The green net is called water-net. If you have found good specimens you will see that the nets are really little closed bags, perhaps like drawing No. 1, Fig. 1, or they may be more slender. Sometimes the bags are several inches long, and the nets coarse, that is, the holes or

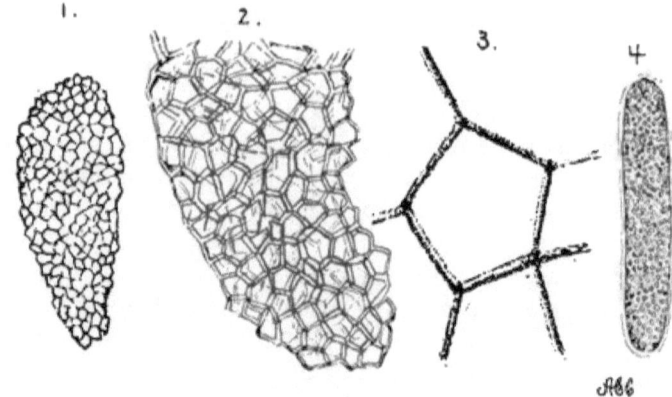

Fig. 1. WATER-NET—*Hydyodictyon.*
1. Colony, x 5.   2. Tip of net, x 15.   3. Single mesh, x 40.   4. One cell, x 75.

meshes are large; but the large nets are easily torn, and are likely to be found in ragged bits. There are usually plenty of smaller nets; if you have sharp eyes you may find some not more than one-eighth of an inch long. Drawing No. 2 is one end of one of these tiny nets as it looks under the microscope, and No. 3 is one of the meshes still more enlarged. Take the coarsest net you have, and find out how many sides the meshes have. How long is the longest side you can find?

No. 4 is a drawing of one side of a mesh under the microscope. So we see that each side of any mesh is itself a little closed bag or sac filled with something. This sac with its contents is called a cell; the sac is transparent, and is called the cell wall. The jelly-like substance within seems to be green, and there are grains in it, some very small, others larger. After water-net has been kept in alcohol for a day or so, the alcohol becomes green and the net is left colorless, but otherwise the cells look just as they did before. This coloring matter, which is dissolved by the alcohol, is a very important part of plants, so important that it is best to learn at once the long name the botanists have given it; it is called chlorophyll. The jelly-like granular substance that still lies close to the cell wall is another very wonderful substance; it is present in every living part of every plant and animal; in your own bodies as well; it is called protoplasm.

Now put some of your water-net in a dish of water in the sun. In a short time you will see little bubbles all about it. If you found the net yourself, growing in a sunny place, you probably noticed that there were so many bubbles entangled with it that the whole mass looked like green froth. It is possible to collect the bubbles that the net gives off, and to find out that they are bubbles of oxygen, a gas that we must have in the air we breathe, in order to live at all.

Dissolved in the water, is another gas that all living things are constantly breathing out; it is called carbonic acid gas. This gas is the most important food material for plants, and these little water plants are drinking it in with the water all day long. Now the cells of the plants are little work-shops or laboratories. The materials used are water and carbonic acid. Each of these materials consists of oxygen united with something else, and in every cell the protoplasm, with the help of chlorophyll and sunlight, breaks up this raw material, uses what it needs to make its food, and gives off the oxygen that is left over into the air for us and for animals generally. The food that the cell manufactures is called organic matter; it is food for animals as well as for plants. The water-net uses some of it at once for its own growth, but some of it is stored for future use in the form of starch. In the picture of the cell, the larger dots represent little stores of starch.

When little water-nets grow up, that is, when each cell becomes as large as it can be, what do you suppose happens next? If there is plenty of food, the protoplasm

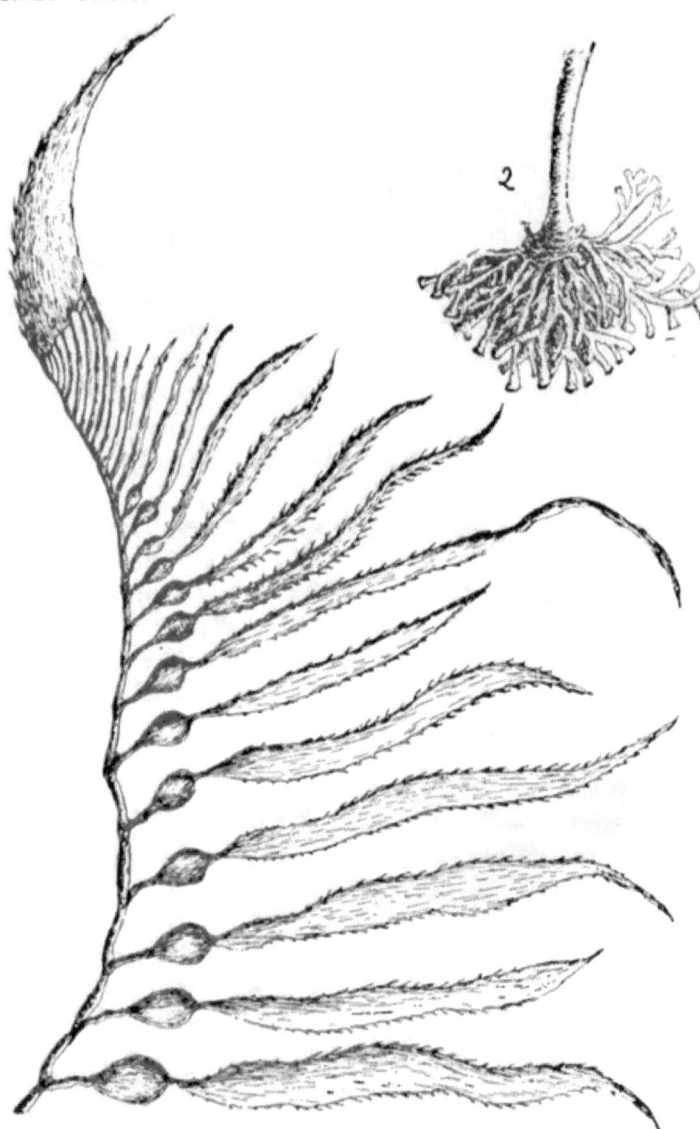

Fig. 2. A BROWN ALGA, or KELP—*Macrocystis*.
1. Tip of branch, ½ natural size. 2. Holdfast of small plant.

in any cell may divide into very many minute particles, into from seven thousand to twenty thousand the botanists tell us. Now each bit of protoplasm, which may be called a protoplast, is able to do a wonderful thing; it can make a room for itself just as a snail can make its own house. So all these seven thousand or more protoplasts enclose themselves in transparent walls, and the old cell wall has within it a multitude of new cells. Can you guess the rest of the story? The tiny new cells move about, arrange themselves into a net, and grow together; the old cell wall breaks, and the tiny new net floats out to begin life by itself. In our sunny summer days the pools are filled up rapidly with new nets; but since our rainless season is so long, the pools may dry up, and then the nets must die. Botanists tell us that, as this danger approaches, the protoplasts into which a cell divides, escape at once into the water and swim about for a time; then two protoplasts from different cells may unite to form a cell that behaves like a seed; that is, one which can rest a long time and can stand drought and heat or cold until it has a good chance to grow into a new plant.

There are many other kinds of green Algæ, as these scums are called. Some of them are very beautiful under the microscope, and many interesting things have been found out about them. You may have a chance to study them yourselves in higher schools.

In the ocean, too, are plants that get all of their food from the water, though not all of them can be said to live easy lives. Often they grow on rocks where the waves are constantly breaking. Perhaps you can imagine, if you have been in the surf yourselves, how these plants are pounded, tossed and twisted by the breakers. The brown Algæ, or rock weeds and kelps, are the most common plants along the shore. Have you ever noticed how they are fastened to the rocks? No. 2, Fig. 2, is a drawing of one kind of holdfast. Sometimes the holdfasts are very large, with many branches. You may have seen them on the beach after a storm. Some are solid instead of branched. If you have tried to pull kelps from the rocks where they are growing, you know that these holdfasts rarely let go. You must either scrape them off with a knife or break the stem; and how tough they are! They seem almost like leather. It is no wonder that they can stand so much beating by the waves.

One kind of kelp common on our coast is called Macrocystis. No. 1, Fig. 2, is a reduced drawing of one end of a branch. Plants several yards long are frequently washed up on our beaches, and travelers tell us that some plants grow a quarter of a mile long. Children often strip off the leaves, and use the stems for skipping-ropes, because they are so strong and flexible. The leaves of this kelp are full of blisters, and at the base of each leaf is a pear-shaped air sac. Can you think out the use to the plant of the blisters and the air sacs? Notice how the new leaves start at the end of the stem. Do leaves of land plants begin in that way?

Fig. 3 is a photograph of one of our red sea mosses. It is often thrown up on the beach during storms, and it is still very pretty after it has been floated out on white paper and dried. Most bright red Algæ grow in rather deep water, and so do not get as much light as plants that float near the surface. The red color changes what light they get, so that is most useful to the plant for food-making.

Of course the ocean plants are in no danger from lack of water. They are sometimes badly torn by storms, but this is not always misfortune. If the pieces happen to reach a favorable place, they grow fast and become new plants. There are crabs that seem to know this fact, for they cut off bits of sea mosses with their claws, and place them on little hooks on their shells. The plants grow there, and hide the crabs from their enemies.

ALGÆ

Fig. 3. A RED ALGA—*Plocamium coccineum.*
Portion of frond, natural size.

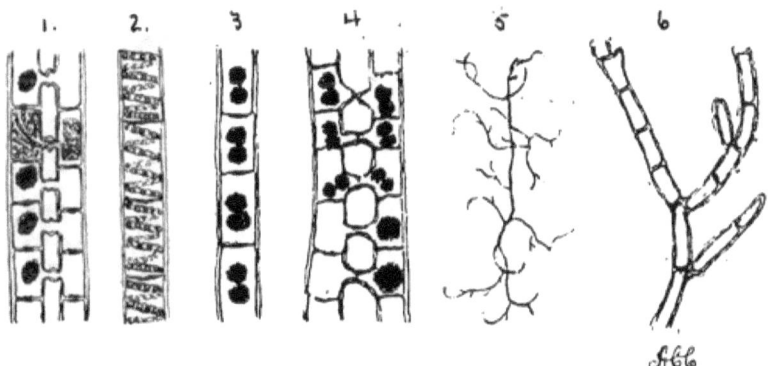

Fig. 4. SOME GREEN ALGÆ.
1. Conjugating filaments of Spirogyra. 2. Fragment of filament of Spirogyra.
3 and 4. Zygnema. 5 and 6. Cladophora. 5, x 4, others, x 50.

You will, perhaps, think that these sea plants have no chlorophyll, and so cannot manufacture food as the water-net does. But the chlorophyll is there in plenty, hidden by the brown or red color. You can see this for yourselves in pieces that have been faded in the sun on the beach, or in pieces you have kept in fresh water. So we see that although water plants get their raw material easily, the pretty red sea mosses and the big kelps, as well as the little water-net, are busy all day long making up this material into useful food substance, at the same time setting free oxygen for the general good. Perhaps it was not quite fair to compare them with the lazy boy.

CHAPTER II.

HOW SOME PLANTS BEGIN LIFE.

Plants that have seeds are called higher plants, because they give their children, the seeds, such a good start in the world.

Take the castor bean for an example. On the home plant, each seed has a room to itself in the little ball of a house. It seems quite safe in the strong walls defended by stiff little prickles; but suddenly, on a sunny day, the walls of the little room split open, sometimes with such force that the seed is flung far out into the world. For a time it lies on the ground exposed, as one might think, to many dangers; really the seed is protected by its hard coat, and by being poisonous. Soon, too, it becomes buried in the dust, where it waits with other seeds for the rain and sunshine to waken it.

Soak some castor beans in water, and you will see how the little knob at one end acts like a sponge in taking up water. Plant the seeds, and in a week or so a little white sprout pushes out where the knob was. Do you think this sprout a root or a stem? Really it is both, as you can see by watching, for the root grows steadily downward, but the stem part grows upward and becomes a long, narrow arch, that lifts the earth above it, and finally breaks through. Meanwhile the rest of the seed swells and bursts its coat. What is now the outside part is a white,

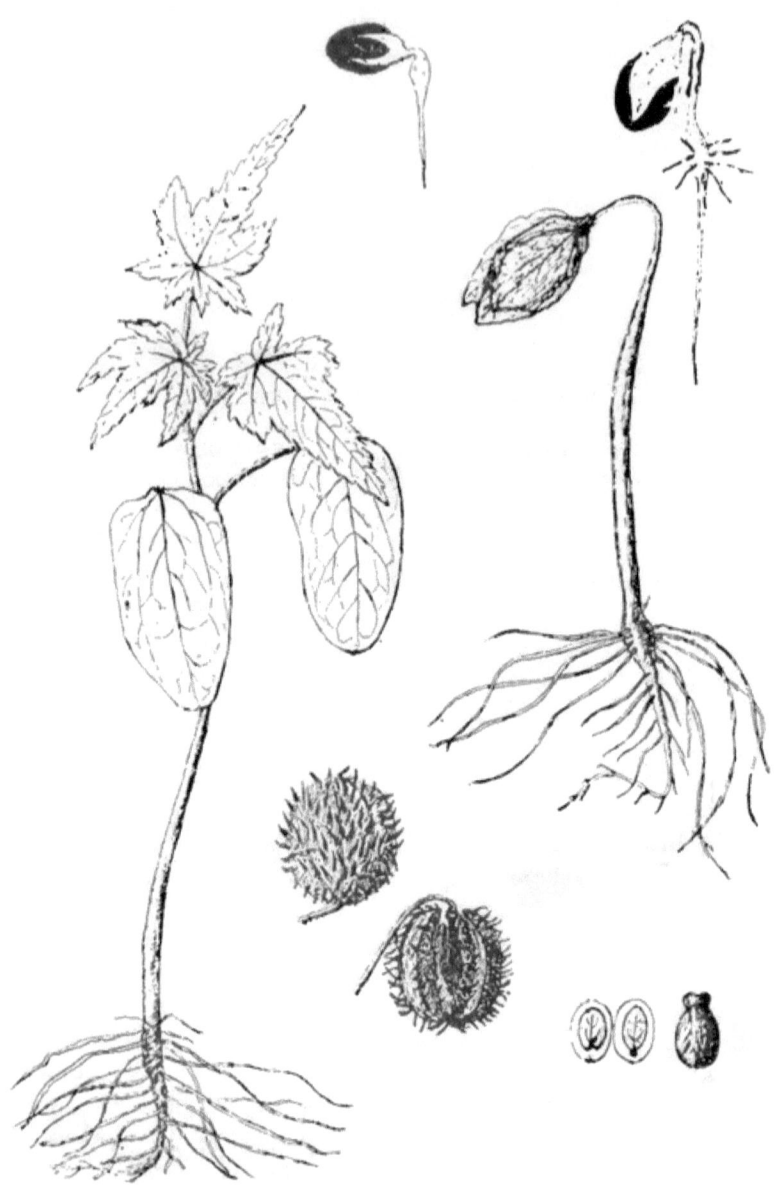

Fig. 5. CASTOR-OIL PLANT—*Ricinus*. Germination of the seed.

oily substance, that feels and cuts a little like cheese; within this are two pale little leaves pressed together. Now take an unsprouted seed, and see if you can find these leaves and the beginnings of the stem and root inside the white substance.

Think out the uses of this white substance. At first it protects the baby plant; later, by swelling, it bursts the coat; but watch it as the seedling grows; the seed-leaves grow larger and thicker, but, at the same time, this substance gets thinner and thinner until it is a mere worthless film. Clearly, this is the food the leaves absorb for their own growth and for the young stem and roots.

Be sure to see how the seed-leaves escape from the soil. For some time only the little arched stem appears above ground. The roots fasten one end of it firmly in the soil, so that as it grows, it must be always pulling at the buried leaves at the other end. Finally the leaves are brought above ground; they turn green and unfold, and now, because they have chlorophyll and light, they begin food-making.

But let us look into the history of a few more seeds. High up in the pine trees, tucked away in the thick-walled chambers, often protected by savage spines, are the pine seeds. The trees seem very unwilling to part with their children, for they sometimes keep them shut up in these pine-cone nurseries for several years before they have done fitting them out for their journey. It really is a journey, for when the cones finally open, each seed has a wing on which it is carried far from the home tree.

Now these seeds are actually in danger, for the food the parent tree has been so long preparing for them is very good for animals too. Squirrels are so eager for pine seeds that they do not always wait to find them on the ground, but climb the trees, and actually cut through the thick, woody walls of the cones with their sharp teeth. But the squirrels bury some of the seeds for future use, and if they forget where they have hidden them, or if they do not live to eat them, the pine seeds are well planted.

Some of our mountain pines have large seeds called pine nuts or piñons. These seeds have enemies with hands as well as enemies with sharp teeth. They are gathered in great quantities by the Indians for winter supplies; they are also sent to our city markets. The piñon has a hard shell, within which an abundance of food is packed about the tiny plant. You will think at first that the baby plant has only a stem, but if you look closely you will find at one end a cluster of very short, thick, white seed-leaves, which can grow into little green needles like other pine leaves. Piñons grow so very slowly that it is better to plant smaller

Fig. 7. GERMINATION OF PINE SEED.

SEEDLINGS

pine seeds that will come above ground, and look like tiny pine trees, in a month or so.

The sprouting pine seeds behave, in many respects, like the castor beans, but the swelling food does not burst the seed coat. Instead of this, the coat is brought above ground by the growing, arched stem, and the leaves remain inside, safe and warm, until they have absorbed all the food. In the picture you can see one way they have of getting rid of their coat. Your seed-leaves may behave differently.

There are many common seeds that you can grow more quickly than the pine and castor bean, such as the morning glory, common bean, squash, pea, peanut, nasturtium, corn, wheat and onion. The morning glory seed has the baby plant and its store of food snugly stored away in the smallest possible space. The common bean has a different plan for storing food; as you remove the seed coat, you see that the two bodies into which the seed separates are attached to the little stem instead of being packed about it. When these bodies are dragged into the light, they become green, and spread out from the stem, like leaves. In fact, they are the seed-leaves just as truly as were the thinner, prettier leaves of the castor bean and morning glory, but the bean seed-leaves have the food which they furnish to the growing plant stored within them instead of about them. They use none of this food for their own growth, but as their supplies are used by the rest of the seedling, they shrivel and finally drop off. By this time, the next two leaves, which were distinct even in the seed, have grown into good food-makers.

The two seed-leaves of the squash also have their food stored inside, but they use part of it for themselves and become very good food-making, or foliage, leaves. Do not miss the clever trick of the squash leaves by which they get rid of the hard outside coat or shell. The stem grows a

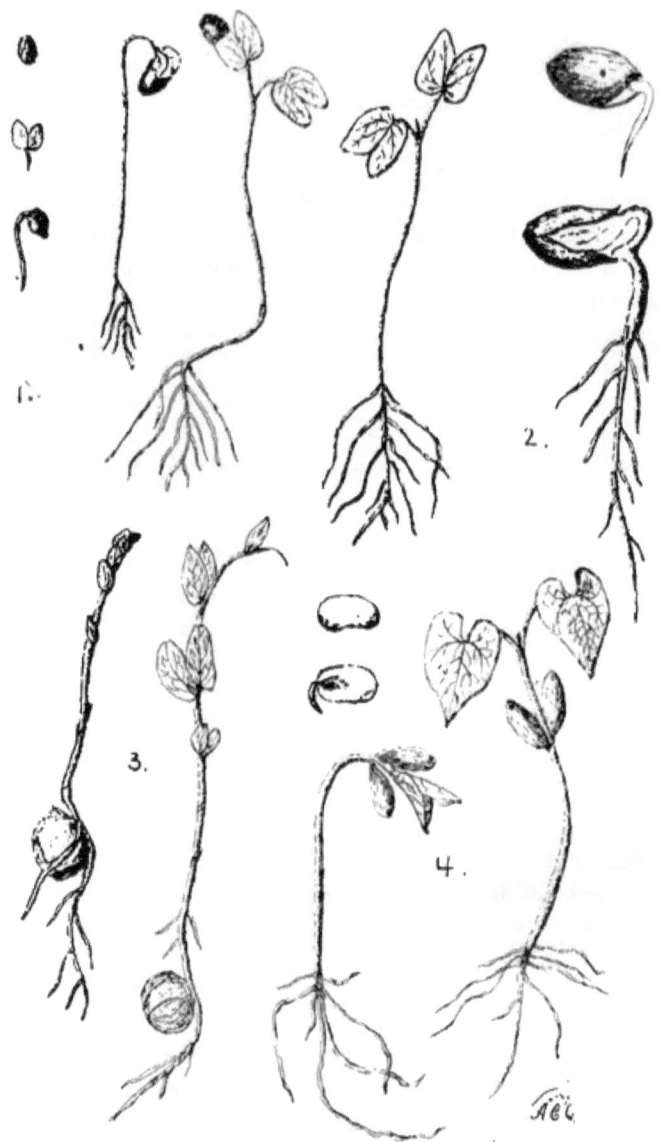

Fig. 7.—GERMINATION OF SOME COMMON DICOTYLEDONOUS SEEDS.
1. Morning glory.   2. Pumpkin.   3. Pea.   4. Common bean.

SEEDLINGS

little knob that pins down one side of the shell, while the arching part of the stem pries up the other side.

The pea has even thicker seed-leaves than the bean, but they do not try at all to imitate foliage leaves; they are content to be simply store-houses and to remain under ground, where they are really better off. This leaves the little shoot between them to push up through the ground alone. Does it push straight up, or appear first as an arch? Which of these seedlings does the nasturtium most resemble?

The peanut? Later in the season keep a sharp lookout for acorns or walnuts that have germinated among the fallen leaves under the trees. In the picture of the sprouting acorn, can you point out the seed-leaves?

Should you like to know the Greek and

Fig. 8.—GERMINATION OF ACORN.

Latin names that botanists have given to the parts of seeds? Seed-leaves are called cotyledons; the little stem in the seed is the caulicle; the beginning root at one end of it is the radicle; the little bud at the other end, plainly seen in the bean, pea and peanut, is called the plumule; the entire baby plant is called the embryo; when food is stored outside of the embryo, it is called endosperm; when the embryo begins to grow, botanists say that the seed is germinating.

The parts of a kernel of corn are not easily made out. The greater part of the kernel is hard and yellow; it shrinks as the seedling grows, and is, clearly, the store of food. Lying closely against this on one side of the kernel is a paler, softer part, and this part is wrapped about a little white rod. When the seed is kept in water for a day or so, you can see that this softer white part swells, and acts like a sponge; that is, it does just what seed-leaves, or cotyledons, always do, it absorbs food for the rest of the seed-plant. So we must consider it a cotyledon, although it does not look at all like a leaf, and it always remains below ground with the endosperm. As the seed germinates, the nature of the little white rod is soon revealed. The upper part shoots upward, and is seen to consist of rolled-up leaves, so it should be called the plumule. The stem part, or caulicle, to which the cotyledon is fastened, does not grow rapidly, but at its lower end the radicle lengthens into a main root; side roots are also sent out. Since the stem does not grow up into an arch, the plumule must push straight up through the soil, but the outside leaf protects the rest, and does not attempt to become a foliage leaf itself. When the stem finally comes above ground, notice the circle of staying roots, like tent-ropes.

The onion seed is very small to study, but it is easy to see that the rolled-up leaves come out as an arch, the tips remaining inside the shell, while the roots grow from the short stem at the base. The onion, like the corn, has one

Fig. 9—GERMINATION OF SOME MONOCOTYLEDONOUS SEEDS.
1. Onion. 2. Wheat, x 5. 3. Indian corn.

cotyledon, which is wrapped about the rest of the embryo, and it is a cotyledon of many uses. It protects the embryo, carries it out of the shell, plants it in the soil, carries it food from the seed's storehouse, and in the meantime becomes a good foliage leaf itself. Palm seeds, too, have single cotyledons which plant their embryos so deep that they are in little danger from drought. One palm has huge seeds weighing fifteen pounds or more; the cotyledon carries the embryo down a foot and a half in the soil, and feeds it there nearly a year before leaving it to care for itself.

As the rains come, keep watch for seedlings that may spring up along the wayside or in fields, vacant lots or gardens. See if they first appear as arches, and if you can find the seed-leaves. As the plumules unfold, see how many of the plants you can recognize.

We have seen how seeds are prepared to begin life by the parent plants; let us think of what the seedlings do for themselves. While the little plant is yet entirely within the seed-coat, it drinks in water from the outside world, and the water dissolves this stored food, and helps to change some of it so that it can be used for growth. As soon as the first white shoot pushes out, it begins to fit itself to supply water. Watch some kernels of corn germinate on moist blotting-paper. The tip of the root is hard and smooth, but back of the tip, the root is covered with dense fuzz, and every hair of this fuzz is a thin-walled cell that eagerly drinks in all the water it can get from the moist air. Roots that grow in the soil have root-hairs too, but the soil clings to them and hides them. They are not usually so long as those you see on the corn, but they are sometimes so dense that the head of a common pin can cover a hundred or more.

Now, as a root grows, it does not lengthen everywhere as you do. If you will mark a corn root with ink, as in the picture, you will see that only the part just back of the tip grows. So the root-hairs that came out first have another use; they unite their part of the root firmly with the soil, and as the growing part lengthens, it, too, sends out hairs to anchor it, as well as to supply water. A great English naturalist, Darwin, found out some wonderful things about the sturdy little root-tips. They are really pioneers and explorers, for they have the power of moving in little circles as they are pushed on by the growing part, and they are also sensitive to hard substances and to moisture; so they are actually able to avoid many dangerous or difficult places, and to pilot the growing root into the very best places for moisture and food.

Take a corn seedling that has been growing for a month or so, remove it from the soil without breaking any of the roots, if possible, and measure the length of all the roots on a string. Now remember, that every root is clothed with root-hairs, and through these thousands of absorbing cells, your seedling has been taking in water. With the water enters what is dissolved in it, and the plants really require much of this dissolved matter. The gas, nitrogen, which is necessary for making protoplasm, is prepared in the soil so that it can be carried up the plant by the water; the soil also supplies necessary minerals, such as iron and sulphur.

But all this raw material must be worked over into organic plant food, and, as we learned from the water-net, the food-making must begin where there is chlorophyll and sunlight, that is, above ground. The path by which this raw material travels up to the workshops, can be easily shown by putting a seedling in water colored with red ink. Soon you will see the red fluid creeping up the woody strands or veins that run through the leaves. By cutting thin slices of root and stem, you will see that the fluid here, too, ascends through woody fibres. You remember that the water-net was made up of cells. So are all plants. The castor-oil seedling, for instance, is really a community of millions of protoplasts, each living in a room made by itself. To see this you must have very thin slices of the plant under the microscope. The little rooms or cells are by no means all alike. To give you some idea of what the cells of the stem are like, two slices, one crosswise and one lengthwise, have been combined in drawing No. 2, Fig. 10,

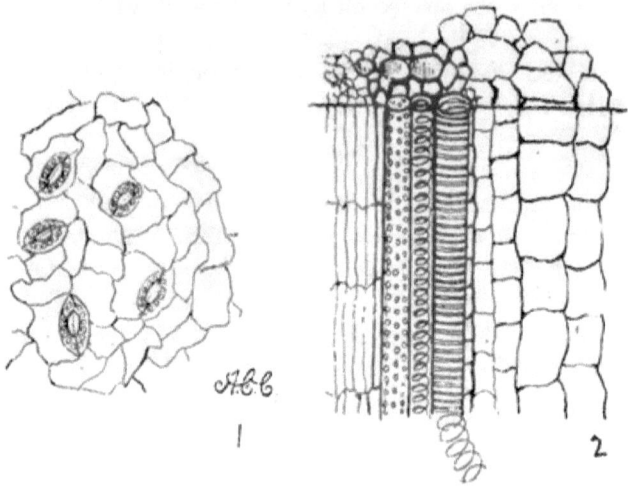

Fig. 10. CASTOR-OIL PLANT—*Ricinus*.
1. Epidermis. x 75. 2. Cellular structure of stem, x 150. (Diagramatic.)

Some cells, you see, are like the water-net cells, with thin walls; others are much more slender and have thicker walls; others are very long and have curious markings on their walls. These last two kinds of cells make up woody fibre like those colored by the ink.

No. 1, in the same drawing is the skin, or epidermis, of the leaf under the microscope. Most of the cells are thin and flat, and fit closely together like tiles, but among them are pairs of cells with openings between them, so that each pair resembles open lips. The openings are called pores, or stomata, and their two guard cells can close together when it is best for the plant. Very little water can pass through the outer cell walls of the epidermis, but when the stomata are open, water given off by any cells within the leaf can pass through the stomata into the air.

Notice that the upper side of the castor-oil leaf is a darker green than the lower. This is because the cells here are packed very closely together, and also because they have a good supply of chlorophyll. The sun shines directly on them during the warmest part of the day. These cells, then, are the workshops, or laboratories, where the food-making begins. The most important material for making organic food is carbonic acid gas. This is taken directly from the air by the cells of the epidermis, and is passed on through their walls to the laboratory cells. The rest of the raw material is brought by the woody fibres. Just how this crude sap is drawn up, is not easy to explain. The sun helps to raise it, and the eager root-hairs and other cells force it up, but not even the wisest students profess to know all about it.

Not nearly all of the water brought to the leaves is used by them. Shut up some of your seedlings in a fruit jar, and in a few moments moisture will be seen on the sides of the jar. This is because the leaves are constantly giving off moisture through their stomata. That is, there is a constant current of water streaming up through plants by way of the woody tissue, and much of the water is used simply to carry the other materials for food-making from the root-hairs to the laboratory cells.

In manufacturing food, the protoplasm, helped by the sunlight and chlorophyll, breaks up carbonic acid and gives off oxygen. The oxygen escapes through the stomata, and you can see bubbles of it if you put plants under water in the sunlight. But this is only a small part of the story of what goes on in the laboratory cells of plants. This first organic food must be further worked up into many, many substances. All our seedlings must make more protoplasm, more cell walls, more chlorophyll, and other colors, perhaps, to paint the flowers with; the castor-oil plant will make one kind of oil, the peanut another, the corn and wheat will make quantities of starch, each plant will manufacture what gives it its peculiar taste or odor, and so on. Did you ever think how many of the things that you eat and wear and use in other ways every day, were originally made by plants? Surely you will admit that these little cell-laboratories are very wonderful shops indeed. This making over of the first organic food into other substances goes on by night as well as by day; often other cells finish what was begun in the leaf cells. The finished products, too, must be distributed to just the places that need them. This is done by quite a different set of cells from those that carried up raw material. So we see that in this little plant-community of millions of cells, each one has its own part to do, and not one is useless or idle.

CHAPTER III.

PLANTS THAT KNOW HOW TO MEET HARD TIMES.

We talk of hard times when money is hard to get, because we want money to pay for food and clothing and shelter. These things the plants make for themselves, but they must have warmth and material to work with. Our California climate is kind to plants in supplying warmth, and the air and soil are constantly furnishing materials that plants require, but oh, how much water they need! Water as part of their food, water for dissolving all other food, and water for carrying food from roots to leaves. In many parts of California, however, there is little or no rain for half the year; so from May to November there are hard times indeed for California plants.

When these hard times come, some plants give up at once, the bur-clover, for instance, the filaree and many grasses. During the warmer rainy months they have lived fast and have made haste to store up food for their many, many seeds, but they themselves do not try to live through the dry season. It is these dead plants that, by the first of June, give California fields and hills their summer tints of brown and gold.

There are other plants that store up food for themselves, usually below ground; as the dry season advances, the leaves which give off so much precious moisture are dropped, and perhaps the shoots die back nearly or quite to the ground; but beneath the soil the plant still lives, although it takes a long summer nap. We shall watch the waking of these plants when the rains come.

Still other plants bravely go right on growing through the long, dry summer. If they live along streams or in shaded cañons, they have not a hard task, for in such places there is water beneath the surface of the soil that the fierce sun cannot quite drink up. So when the rest of the country seems brown and bare, we have along stream beds, lines of willows and cottonwoods, alders with their clean, straight trunks, great branching sycamores, and perhaps smaller plants with pretty flowers. In our cañons and on steep, shaded slopes we have a great many plants that grow and flourish throughout the summer. The hardy nightshade and the wild broom go on flowering as they do in the other seasons, year after year. The poison oak is able to keep its beautiful glossy leaves; the California holly blossoms and gets ready to ripen its berries for Christmas; the clematis carries its fluffy seeds high up where they can get a good start in the world, and the grape vines spread out their leaves to the light.

But there are less sheltered hills and open fields and waysides on which the sun beats all the long, cloudless days; these places, too, have some plants that do not rest during the summer and autumn. Collect all such plants that you can find on a September or October day. Get the roots when you can; at least dig for them until you know something about their length. Now try to describe your collection. It will not contain a single bright green plant. The leaves are all dull green or grey; sometimes because they have hard, thick skins, more often because they have hairy coats. These coats are not fine, silky or velvety ones, such as young leaves often wear in spring-time; they may be fleecy, or they may feel like felt or flannel, or the hairs may be very short, but they are rough and coarse; often, too, they are sticky or prickly and are very disagreeable. Some of your plants will have small leaves, or else few leaves; and you may find some with no leaves at all. The stems may be hairy like the leaves, and are likely to be hard and woody. If you collect your plants along the sea shore, or in a very dry or sandy region, you will have some with very thick, fleshy leaves or stems. Find out what you can about the taste of your plants. Are any of them poisonous? Do animals eat them? You are sure to notice the odors. You could learn to know many California autumn plants by the sense of smell alone.

Now we want to think out the uses of these qualities. What dangers do they help the plant to meet? At this time of the year, when plants are so scarce, they are in special danger of being eaten by insects and grazing animals. So it is easy to see the use of poisonous or harmful qualities, of bitter taste or disagreeable milky juice. The strong odors, of course, give warning of unpleasant taste. Sticky and hairy leaves are not pleasant to the tongue, and sometimes prickles are very cruel weapons indeed. At the end of this chapter are drawings of some common plant hairs under the microscope. Number 6 is one of the little bristles of our common cactus, or prickly pear. Do you wonder that it is so much trouble to get them out of your flesh? It is not strange that animals learn to avoid the plants.

Another danger that summer brings to plants in the drier parts of the state, is the intense light, for it injures the chlorophyll unless a screen is provided. So here is another use for the thick skins and hairs that give the plants their gray color. Perhaps you know how gray our desert plants are. Another danger is dust. Plants, like animals, breathe and can be suffocated. The hairs of plants help to keep the dust from their pores, just as your lashes protect your eyes.

But the greatest danger that threatens these plants is that the roots may not be able to supply as much water as the parts above ground give off. When you shut up corn leaves in a glass jar, the moisture they gave off could be seen in a few minutes. In the open air evaporation is more rapid. Some one has figured out the amount of water given off by an acre of grass in twelve hours. It is more than one hundred tons; it would cover your schoolroom floor to the depth of three or four feet. Surely, our dry weather plants cannot afford to part with water at this rate.

Find out for yourselves if the plants you have collected lose water as fast as the corn. Shut up in separate jars equal weights of corn seedlings, and of shoots like hoarhound or tarweed with woolly leaves, twigs of live oak or Eucalyptus with hard leaves, and a fleshy plant like a cactus or a Sedum. Do they give off different amounts of water? Take off the covers and leave the plants exposed to dry air and the sun for a day, then weigh them again, and see which have lost the most water. Next find out how much the skin helps to keep in moisture. Peel off the skin from some leaves, Sedum leaves are good for this; put these leaves and an equal weight of entire leaves in the sun for a day or two, then weigh them again.

It is very clear now that the skin of plants control the giving off of water, and that a hard or hairy skin keeps in more than a delicate one, such as the corn has. Turn back to the picture of the epidermis, or skin of a leaf, in Fig. 10. It consists, you remember, of tile-like cells fitted closely together, and of pairs of cells with openings between them, the pores, or stomata. Now the tile-like cells have their outside walls thickened, sometimes very much, as in the live oak, Eucalyptus and cactus. This thickened layer contains a substance somewhat like tallow, through which it is very difficult for water to pass. So most of the water leaves a plant through the pores. But these little mouth-like pores can open and close; when the air is very dry they close and so help the plant, for it is of course when the air is hot and dry that the plants are in greatest danger. Still all these devices do not entirely prevent evaporation. It would be bad for the plants if they did; for if the current of water from roots to leaves were to stop, there would be no raw material brought from the soil to the leaves, and food-making would soon come to an end. So even hard leaves, like the live oak and Eucalyptus give off so much water that in our experiments we can see and weigh it. We use only a few leaves in our experiments; think how much water a great oak or a tall Eucalyptus must give off from all its leaves on a hot summer day!

The live oak, even in Southern California, will grow a long distance from streams, but it has many roots, some of which go far and deep in their search for moist soil. So the oak does not rest entirely during the summer, but keeps busy enough to ripen its acorns. The Eucalyptus lives much faster. It sends out new leaves all summer long, even when it grows in places too dry for the live oak to live at all. Its home is Australia, and it seems to have learned there how to defy drought. You know what greedy roots it has; how they drink in all the moisture from the ground so that other plants that try to live near it starve. You have noticed how little shade the leaves cast at noon. This is because they have their edges toward the sky, and the sun cannot strike their surfaces during the hottest part of the day, and so draw out more moisture than they can spare. Earlier and later in the day the sun shines directly on the blades, and helps them to make their food rapidly. Other trees that have been brought from foreign countries to California, because they know how to stand dry weather, have this same habit of vertical leaves. Find out some of them yourselves. Our own Manzanita, too, has this habit.

In your experiments you found that the cactus and the Sedum lost less water than any of your other plants. It is easy to see why. If you were to take two equal lumps of clay, and roll one up into a ball and make the other into thin leaves, you know which would dry out first. Indeed the cactus know so well how to store water that it can live through the summer in deserts where all other plants perish entirely or die back to the ground.

Other plants beside the cactus adopt the plan of presenting little surface for evaporation during the dry season. Your collection is sure to include plants with few leaves or with small leaves. The wild broom, Fig. 11, has both small and scattered leaves, but the stems are green, too, and help in the food-making, so that in unexposed places this plant can go on flowering all the year round. The plant in Fig. 12 is one kind of Eriogonum. There are nearly two hundred kinds, and they abound in the drier parts of Western America, often where few other plants can exist. The Eriogonums have usually small, woolly leaves. Often the leaves grow close to the ground and die early in the dry season; after this the plant does not grow much, but it can develop flowers and fruit by means of food already stored, helped out by what the green cells in the woolly stems manufacture.

The plant pictured in Fig. 13 belongs to the same family as the tarweeds, sage brushes, everlasting flowers, sun flowers and many others that you have probably collected. This family is called Compositæ; and it is the very largest and perhaps the most clever of all plant families. We shall study the flowers in the spring, but you will be interested then to remember how many of the family know how to meet hard times. Their devices are many, as you have seen, but the most common one, perhaps, is a hairy covering.

A thick coating of hairs is probably the best protection of all for plants that are really active during the summer.

AUTUMN PLANTS

Fig. 11. WILD BROOM—*Lotus glaber.*   Fig. 12. *Eriogonum elongatum.*

The everlasting plants, the hoarhound, and the sage brushes are good examples of this. The hairs are empty, or rather they contain simply air, so they keep in moisture just as a layer of straw keeps soil moist, or as a stopper of cotton in a bottle keeps water from evaporating. Besides, the interlacing hairs actually keep the leaf cells cooler by shading them from the sun. At the end of this chapter are some of these screening plant hairs as they look under the microscope. One of the most hairy of California autumn plants is the one in Fig. 14; children sometimes call it turkey-weed. In some parts of the state it covers acres of hard, sun-baked soil, but it is so nearly the color of the dusty earth that it is easily overlooked. It is sticky, prickly and ill-scented, and you may think it a very ugly little weed. Indeed, with so many dangers to combat, plants cannot always afford to be pretty, they must be content with being clever. But really this turkey-weed is not without beauty. No. 4 at the end of the chapter is a drawing of the hairs under the microscope. You see that they are little stars, each having

Fig. 13. A SUMMER COMPOSITE—
Malacothrix tenuifolia.

one long point that serves as a dagger. Do you not see what an elegant screen these interlacing stars form against intense light, heat and dust? But these same hairs have still another mission. Have you ever picked the plants on a dewy or foggy morning and noticed how heavy they are? You can find by experiment that the leaves take

Fig. 14. TURKEY-WEED -*Eremocarpus setigera*.

in their own weight in water in a few hours, and under the microscope you can see the water enter the hairs. Probably this is why the plant can go on blossoming and ripening seed after it seems impossible for the roots to get moisture from the soil.

No. 2, Fig. 15, is another disagreeable weed very common in Southern California, but its flowers are pretty and very interesting indeed. They keep honey for the bees only, but they know how to make the bees pay for it by rubbing pollen against their stigmas. No. 1 is the wild fuchsia, common on many California hillsides; and a very beautiful plant it is! It is not strange that the humming birds like to visit its brilliant scarlet flowers. If you watch, you can see how the birds carry pollen for the flowers.

Perhaps, as you were collecting your plants, you noticed that the poison oak leaves were turning red, or that the sycamore leaves were beginning to look dull and dry. Really these plants were getting ready to take a rest, perhaps because the summer had been so hard for them. In cold and in dry climates there are many trees and shrubs that do not try to meet hard times, but drop all their leaves and remain idle until better times; they are called deciduous. In very dry countries, the leaves fall when the dry air takes too much moisture from them; in cold countries, when the soil gets too cold to supply water. It is not easy to explain why our trees and shrubs drop their leaves just when they do. Most plants have definite times for doing certain things, and sometimes it is impossible to explain why; then we give the fact a long name, we call it periodicity.

Perhaps it seems to you a great waste for the trees to lose their thousands of leaves every year; but the waste is not so great as it seems. Before the leaves fall, the most valuable materials emigrate to the stem, to be used later by the new leaves. The chlorophyll is broken up and carried away, and, in the many changes that take place, sometimes very brilliant colors are produced. Perhaps you know something about the gorgeous autumn colors of forests in our Eastern States. Old leaves always contain much mineral matter that has been left over from food-making. Have you not noticed the amount of ashes remaining after leaves have been burned? This waste matter is given back to the earth as the leaves decay, and can be used again. The decaying leaves enrich the soil in other ways, and also help to keep it moist.

Be sure, during the rest of the year, to watch the

Fig. 15. PLANTS OF THE DRY SEASON.
1. Wild fuchsia, *Zauschneria Californica*. 2. Blue curls, *Trichostema lanceolatum*.

CALIFORNIA PLANTS IN THEIR HOMES

plants you have been studying. You will find some of them working on more busily than ever after the rains come. Others will take their vacation just when material for food-making comes most easily; but you must remember that in plant life, as in business life, there is what is called competition. The stronger crowd out the weaker, and there is not room for all at the same time. So the plants that know how to meet hard times wait until there is less competition, and brave other dangers instead; and, as we have seen, they make a success of life and are sometimes beautiful besides.

Fig. 16. PLANT HAIRS UNDER THE MICROSCOPE.
1. Croton.   2. Mentzelia.   3. Gnaphalium.   4. Eremocarpus.   5. Trichostema.
6. Cactus.

CHAPTER IV.

SOME PLANTS THAT DO NOT MAKE THEIR OWN LIVING.

The world is full of vagabond plants, and the smallest ones make the most trouble. Many of these plants belong to the group bacteria; they are also commonly called microbes or germs. They are very small indeed, so minute that it takes skillful hands and the best of microscopes to find out about them. But they are everywhere, in the air, in the soil, in water and in the tissues of plants and animals. Like all other plants considered in this chapter, bacteria have no chlorophyll, and so they must live on other plants and animals, living or dead. When these little plants have warmth and moisture and just the right sort of food, they grow and multiply at a most astonishing rate; many millions of plants can arise from one, in a single day. When they cannot get food, they simply rest and wait. Some kinds can remain dormant for years, and they are so minute that they can be scattered far and wide in this condition.

In order to get their food, bacteria must break up organic substances and cause many changes. Some that live in the blood of animals cause serious diseases. Often a knowledge of the habits of these tiny plants helps us to get rid of them. It is known, for instance, that a few hours of sunlight and dry air will kill the bacteria that cause consumption, and that boiling temperature kills many dangerous germs. In general, a study of these plants has proved that cleanliness and care can prevent many diseases.

But even bacteria are not all bad. When plants and animals die, it is the bacteria that cause their decay; that is, this multitude of invisible plants, in getting their own food from dead organic matter, break it up. In taking what they need for their own growth, they set free valuable substances that would otherwise remain locked up. Carbonic acid gas is returned to the air, and the nitrogen that formed part of the dead plants and animals, is prepared by the bacteria so that it can be used by living plants. So finally all the dead matter is used up and disappears, leaving room in the world for new generations of plants and animals.

There are other invisible plants, almost as small as the bacteria, that make our bread light and edible. When we stir tiny yeast plants into flour and water and keep them warm, they become active at once, and eagerly help themselves to the food that the wheat made for its own seedlings. Bread is raised by the bubbles of gas that the yeast plants set free, as they take what they want for themselves. It should be baked in time to kill the little plants before they rob it of so much food that it is sour. In like manner minute plants cause what we call alcoholic fermentation in beer and in wine and fruit juice generally. When fruit is canned it is heated in order to kill all such mischief-makers.

But there are also vagabond plants that we can see without the microscope. Keep some bread moist and warm for a few days and watch it closely. You will see first a mass of fuzzy, white hairs; soon you will be able to make out clusters of tiny white stalks with balls on the ends; finally you will see the balls turn black and the whole mass darken. Your bread will then be covered with a ripened crop of plants called black mould. One of the clusters under the microscope looks like No. 1, Fig. 17. You see there are root-like cells for sucking in the ready-made

FUNGI

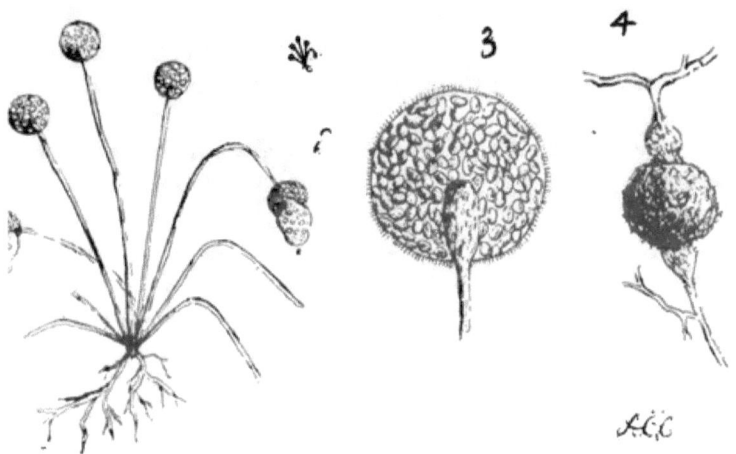

Fig. 17. MOULD.
1. Single plant, x 40. 2, same, natural size. 3. Spore-case. 4. Resting spore.
3 and 4 highly magnified.

food. You probably guess that the minute bodies escaping from the balls correspond to seeds and are able to grow into new plants; they are called spores. Think how many spores every little plant produces! Since there are so many mould spores in the air, do you wonder that whenever we leave their food standing long enough in a warm, moist place, we are sure to find a crop of mould?

There are many other plants that, like the mould, consist mainly of slender, white hairs, but the hairs are often densely interwoven. Fungi is the name given to this group of plants. When fungi get their food from living plants and animals, these delicate hairs, or cells, that absorb the food are often within the tissues of their host, but the spores are likely to be produced outside. Smut on corn, rust on wheat or rose leaves or malva, and the white powder often seen surrounding dead house-flies, are all spores of such fungi. There are fungi that sometimes attack potatoes, grape vines, fruit trees, silk worms, etc., and work great ruin. There are men who spend their lives studying these fungi under the microscope. Often they are able to find out how to destroy them, and then millions of dollars are saved.

Lichens do not deserve to be classed with vagabond plants, as we shall see. You will usually have to look in moist places for lichens. It is said that Indians find their way through forests by means of the denser growth of lichens and mosses on the north sides of trees. Lichen No. 1, Fig. 18, grows in long gray festoons from trees along our coasts, and a very ancient, weird appearance it gives them. No. 2 is a dainty little lichen that clothes shaded banks of crumbling rock; in fact, the lichens are constantly crumbling this gravel into finer soil, because the white threads that fasten them to the rocks can actually dissolve the rock and pry fragments apart. No. 4 is very abundant on weather-beaten pines in the mountains; the color of the lichen is a soft, greenish yellow and its cups are a rich brown. No. 3 is one of the many lichens that form crusts on rocks or old wood. No. 5 is a dead twig on which several lichen colonies have found a home. It is not rare for dead trees, great rocks, old fences, walls, buildings, etc., to become really beautiful because of a covering of lichens with their graceful outlines and beautiful tints. The coloring matter of some lichens is used for dyes.

Lichens grow the world over. They have been found on the highest mountain tops ever reached by man; they grow in hot countries on rocks that are so heated in the dry season that you could not bear your hands on them; they also thrive in arctic regions. The so-called reindeer moss is really a kind of lichen, and this lichen makes life possible, not only to the reindeer, but also to the people that depend on the reindeer for their living.

Make a collection of lichens, and find out all you can about them. As you pull them off from whatever they are

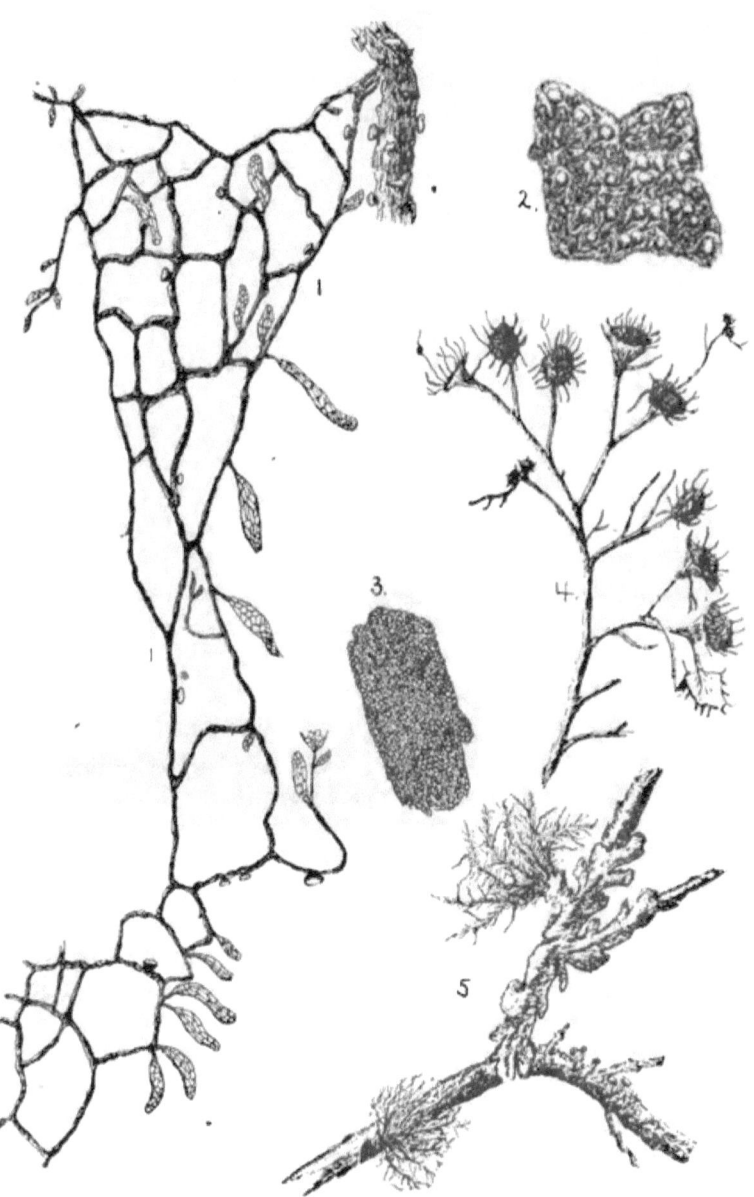

Fig. 18. LICHENS.

growing on, you will notice that you must break tough, white fibres that have held them fast. Perhaps, by tearing and picking apart your lichen, you can find out that much of it consists of this same tough, white, felt-like mass of fuzzy hairs; that is, the greater part of a lichen is fungus. Probably you can find near the surface some traces of bright green color. You will find little cups or saucers on most of your lichens. You can see, too, how quickly dry lichens absorb nearly their own weight of water.

Now the microscope tells a wonderful story about lichens. It shows clearly that the greater part of the lichen is fungus, but among the interlacing threads of the fungus are multitudes of tiny, green cells, called algæ. Each green cell is a plant that can live quite by itself. So we see that a lichen is not a single plant, but it consists of a big fungus that cannot make its own living and a host of minute algæ that can make food very fast indeed when there is plenty of water; and the big, helpless fungus is actually supported by these tiny mites, the algæ. But the algæ can well afford to make food for the fungus, as well as for themselves, for the fungus shelters them and provides them with the moisture and the dissolved minerals and gases that they need. The algæ really thrive better imprisoned by the fungus than when they are free and have only themselves to support. So both fungus and algæ have plenty of food when there is moisture to be had, and they can get moisture from the air when we cannot see it at all. When the air is dry, they rest; and they can rest for months, and then spring quickly into active life again if there is rain or fog or dew.

Like other fungi, the lichen fungus produces many spores. They grow in tiny sacs just below the surface of the little cups you saw. When these spores find a resting place in a favorable spot and among suitable algæ servants, they grow into new lichens. You will often find lichens covered with gray powder. This consists simply of lichen fragments, which also can grow into new plants.

Another group of attractive and often useful fungi includes toad-stools, puff-balls, earth-stars, shelf fungi, and others. Most of these grow in decaying plant or animal matter. When the plants grow in decaying leaves, the part that takes in the food is easily seen. Sometimes it looks like mould, but usually it is more compact. This part of toad-stools is sometimes surprisingly small, but the slender, delicate cells can take in food and make new cells at a most marvelous rate. Have you not seen great masses of toad-stools come up within a few hours after a rain? And the more you examine the toad-stools, the more you will wonder that they can grow so quickly. The hundreds of folds or gills that hang under the umbrella-like part are like velvet. Under the microscope, we can see that they are densely covered with short hairs, or stalks and every stalk bears four spores on the end. Leave the umbrella, gills downward, on a piece of paper, and soon you will see the spores that have fallen like powder on the paper. Toad-stools have clever ways for scattering their spores. Many, as they ripen, dissolve and spread over the ground like thick, black ink, so carrying their spores some distance. Many toad-stools invite flies and beetles to lay their eggs in them, promising good food for the little larvæ as they hatch out—worms or maggots perhaps you have called them. The larvæ eat greedily and grow very fast, and soon the whole toad-stool is a wriggling, squirming mass; so the larvæ are covered with spores, and when they burrow into the earth to change into flies and beetles, they carry these spores with them.

The toad-stools, which are good food for baby flies and beetles, are often good for us. They are really about as valuable food as meat, and in countries where the people know and appreciate them, the fields and woods are eagerly

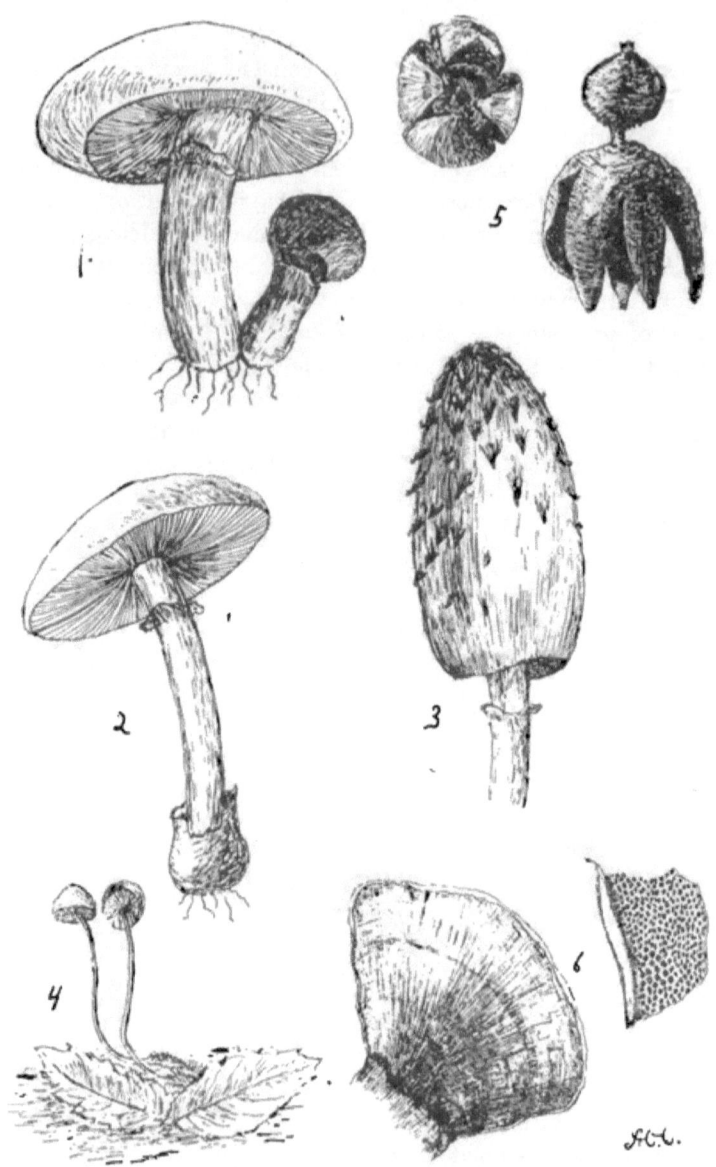

Fig. 19. SOME COMMON FUNGI.
1, 2, 3, 4. Toad-stools. 5. Earth-star. 6. Pore fungus, *Polyporus*.

searched for them. Perhaps you have thought that there is only one kind good to eat, the kind usually called the mushroom, No. 1 in the picture. Really there are many edible kinds. In our Eastern States over one hundred kinds are known to be good for food. A botanist in Southern California found twenty-seven edible kinds in one season, and only two that were poisonous. There are a few kinds that contain a deadly poison. Many others are harmless but are not pleasant to the taste. There is no sure rule that distinguishes all poisonous toad-stools from those that are edible, but people can learn to know them just as they learn to know flowers or birds.

Puff-balls keep their spores inside the balls; pinch a dry one and you will see the spores come out in little clouds. If you have sharp eyes you may find earth-stars among decaying oak leaves. As you can see from the picture, they are similar to puff-balls, but have an extra coat that splits and opens out like a star when the ground is moist; as the earth-star dries, the coat closes up around the spore-case again. One kind of puff-ball grows very large, a foot or even a yard through, it is said. This giant puff-ball is very good to eat. Think how many slices of puff-ball steak can be cut from one of them! The shelf fungus, or pore fungus, No. 6, usually grows on trees, living or dead. The spores line the tiny tubes on the under side of the shelf. The part that takes in food threads through and through the wood, and causes it to crumble and decay rapidly.

There are also plants with flowers and seeds that cannot make their own living; the dodder, or gold thread, for instance, which twines so closely around other plants, and sucks out their juices. On the ground in our mountain forests are numbers of flowering plants with never a trace of green color about them. Perhaps you have seen our beautiful crimson snow plant with flowers that humming birds love. In dense tropical forests, plants of this sort are very common, for, since they do not need the sun to help make their food, they can live in dark, gloomy places. The very largest flower known grows on such a plant in the forests of Sumatra. The flower is more than three feet across, and its seeds are scattered by elephants.

CHAPTER V.

AFTER THE RAINS.

How and when did they come up, these wayside seedlings? Did you watch? Did you catch the pale little arches lifting the soil? Or had the little pairs of leaves got out before you knew the seeds were awake? You were alert if you saw it all, for the seedlings come very quickly at the call of the first rain. Of course you were not surprised to see that the first two leaves were unlike the next ones, for you would remember the seed leaves of the plants you grew in the house. Did you find out that the pairs of leaves that look like little, green, Indian arrow-heads belong to that common weed, the malva? The filaree is easy to make out; the grass comes up like the corn and leaves its cotyledon in the ground; the bur-clover seeds stay in the bur and sprout there. And this is a very good plan the bur-clover has. Germinate some bur-clover seeds on top of moist earth, and you will see what trouble they have getting into the soil. The root-tip wants to go down, as root-tips always do, but the seed is very light and it is not held down by the earth above it, so the growing root has nothing to push against, and simply slips along the surface until its root-hairs anchor it so that it can bore its way into the soil. But the burs that hold the clover seeds are more or less covered with dust; when it rains they stick in the wet soil, and their little teeth anchor the seeds so firmly that

Fig. 20. EARLY SEEDLINGS.
1. Fox-tail grass. 2. Malva. 3. Filaree. 4. Bur-clover.

the root can grow down at once. And the seeds are better off in the bur than they would be in the ground, because they are protected and, at the same time, are likely to be carried away in the bur to some place where they will have more room to grow.

Perhaps, after the first rain, there were weeks of hot, dry weather, and the little seedlings that came up first died of thirst; but nature seems always to have a reserve supply, buried deeper, perhaps, and sooner or later the hills and fields have their carpet of green. Now the seedlings that form this carpet have but a few months to live, and their lives, which seem so short to us, must be very busy ones indeed. There is usually plenty of water at this time of year, and the sun is no longer to be feared, but has become a genial friend. And how the little plants reach out to him! Notice the malva leaves in the morning, and again at noon and towards evening. All day long the leaves turn on their stems so as to directly face the sun, for the sun's rays furnish the power for the food-making that is going on in the millions of laboratory cells packed so closely just beneath the upper surface of the leaf. The bur-clover, too, holds up its leaves to the sun. One kind of filaree, when it has room, spreads out flat in pretty leaf rosettes,

WINTER PLANTS

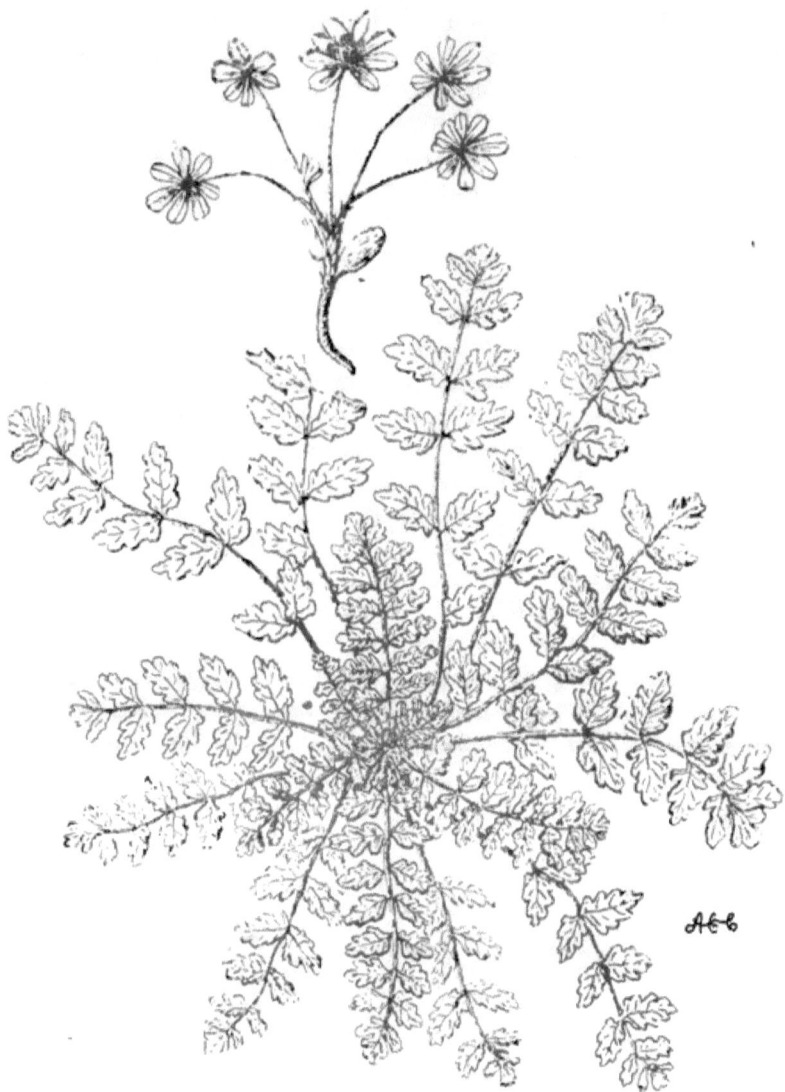

Fig. 21. SEEDLINGS SPREAD OUT TO THE SUN.
1. Lupine. 2. Filaree.

as in Fig. 21, but when many plants grow close together, the leaves rise and stretch up to the sun. Count the leaves on such a plant and see how many are not shaded. Even if one leaf must stand between another and the sun, it is so slashed and divided that it lets much light through. The lupine seedling in the picture spreads out like a fan with leaves facing the sun; it looks almost as if it had been pressed for a herbarium. This lupine has a coat of silky hairs that later on will preserve it from loss of moisture, but now it protects it from the cold of the nights. The bur-clover leaves know a good way to keep warm at night; they fold their leaflets and cuddle them close together so that they lose less heat. You can see this easily for yourselves, for the clover leaves go to sleep early. Try to find other leaves that take "sleeping positions."

But our green carpet does not consist entirely of seedlings. Soon after the first heavy rains, plants like those in Fig. 22 appear and soon overtop the others. By digging, you can find out why they grow so much faster than the seedlings. No. 4, Fig. 22, is the soap-root, cut in two to show what the bulb is like inside. The soap-root is very common in California, and it is well named, for it does very nicely for soap. Examine a plant carefully and you will find that the white layers are the bases of the leaves, and that the layers of brown husk are the remains of leaves of other years. The white substance is, of course, mainly food, and the soapy quality protects it against hungry little gophers and the like; the husk of strong interlacing strands is a protection too. Compare the amount of food in this bulb with what is stored in little seeds like the bur-clover or malva, and you will understand why the soap-root can grow so fast. As long as there is plenty of moisture, its pretty ruffled leaves go on making more food and storing it in their bases; but do you think these leaves can go on working when the dry season comes? They are like the

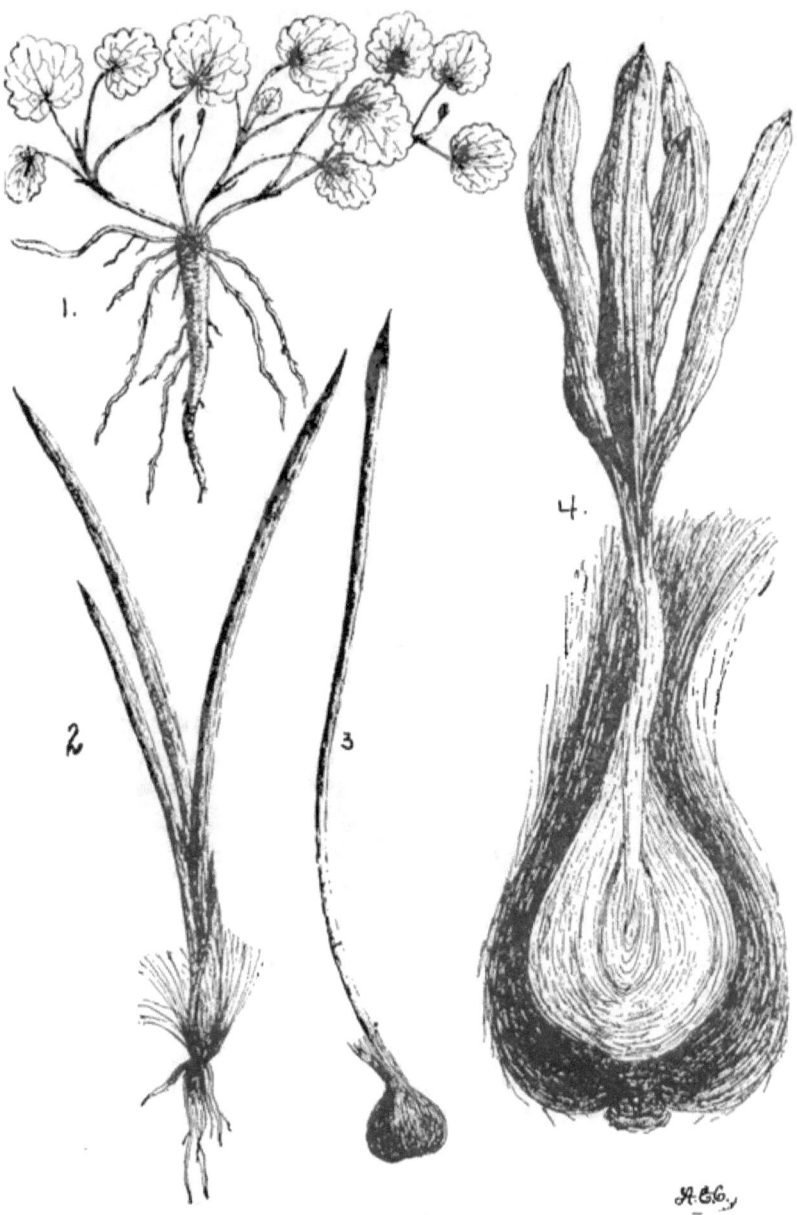

Fig. 22. EARLY GROWTH FROM UNDERGROUND STOREHOUSES.
1. An Umbellifer. 2. Blue-eyed grass. 3. Cluster lily. 4. Soap-root.

corn leaves, and give off moisture rapidly, and when the soil gets dry they are sure to perish. It is not until after they are dead, that the food they have stored will be used. In June, perhaps, a tall flower stalk, sometimes six or eight feet high, will shoot up from the bulb, and slender, white lilies will open late in the afternoon,—but we will wait till June to see the rest.

No. 3 is one of the earliest plants to bloom. It is the cluster lily, cocometa, the Spanish call it. We shall find out more about it in Chapter VII, but now is the time to see what the first leaf does; it has pierced the ground by means of its sharply pointed tip and it is wrapped about the other leaves and the beginnings of the flowers, so that it brings them, too, above ground and protects them for a time. No. 2 is a young plant of the blue-eyed grass. No. 1 belongs to the carrot family, and will bloom early. You will find that many other plants that were entirely underground during the dry season, are starting into life again now. The prettiest of them all are the ferns, but they shall have a chapter to themselves.

A very striking new growth from an underground store-house is the chilicothe, or wild cucumber. The shoots come up early in the winter and in a very short time they are many feet long and are in full flower. In the valleys of Southern California one can always find the chilicothe in flower before Christmas, even in dry years when there is very little other new growth at that time. The root explains this, in fact the plant is often called big-root or man-root; roots two feet long and half as thick are common, and they are said to be sometimes four feet long. These roots must store up a great deal of moisture as well as food, for when the regular season for new growth comes, they can supply the shoots regardless of rainfall; in fact, new shoots will come out in November from roots that have been out of the soil for months. Of course such roots

Fig. 23. CHILICOTHE—*Micrampelis macrocarpa*.

as these are very troublesome, greedy neighbors in orchards, and they grow so deep in the soil that it is not easy to get rid of them. The roots protect themselves against animals by being very bitter. The foliage is bitter, too, but cows will eat it; perhaps you know how it makes the milk taste. The vines climb up into the light by tendrils that grasp everything within their reach, and the flowers burst out before the leaves are fully grown. There are two kinds of flowers; see if you can find them. The flower that will become a big, spiny, green bur, grows by itself close to the stem; you will find the little bur below the white corolla; above it is a big sticky knob called the stigma. The flowers that grow in clusters have, in the center, organs that produce yellow dust called pollen. The pollen must reach the stigma in order to make the bur and its seeds grow. As these flowers do not make honey to induce insects to visit them and to carry pollen to the stigmas of other flowers, the pollen is probably carried mainly by the wind. You must watch the burs after they are grown, to see how the seeds get out. During the summer the softer part of the bur decays and leaves a pretty lace-like skeleton.

Look now for the plants you studied in the fall. Many of them are taking their winter rest, but the hoarhound, everlasting plant, and sage brush have fine new shoots in fleecy, white dress. Some shrubby lupines are even more beautifully dressed, in shimmering silky coats. The poison oak has already its new leaves, and see how they spread out to the sun! When the poison oak is climbing against a bank or tree trunk, the leaves fit so closely together without overlapping that they form what are called leaf-mosaics. How does the poison oak climb? Does the chilicothe climb in the same way? What cultivated plant does?

Many other shrubs that lost their leaves in summer or autumn, have their new leaves well grown, and some are in flower, the wild currants and gooseberries, for instance.

Fig. 21. WILD CURRANT—*Ribes glutinosum.*

The leaves of the wild currant are rather disagreeable; they come early and have to look out for themselves, but the pink and white flowers are lovely enough. Eastern people say they resemble the trailing arbutus, their own early spring flower that they love to talk and write about, and they wonder that we do not care more for our pretty early flowers. Some early wild gooseberries have long, bright red flowers that serve honey to humming bird guests, but do not exclude the larger bees. The gooseberries are well protected by thorns, so they can afford beautiful glossy leaves; they arrange their leaves so that they get plenty of light and at the same time form a roof over

Fig. 25. WILD GOOSEBERRY—*Ribes amarum*.

the flowers that will protect the pollen from rain. Perhaps you will find the wild lilac in flower. One kind of early lilac has leaves with the edges rolled back all around and a thick fur on the under side. In the cañons you will find many other plants with rolled leaves, or leaves with fur or fuzz or bloom on the underside only, and we shall talk about the reason later on.

How many of the native trees of your neighborhood are leafless now? Notice the cultivated ones, too. Have any of them new leaves yet? Trees are generally slower than smaller plants in getting their new leaves, and most of the deciduous trees will seem to you quite asleep this time of the year. But you can find the buds that will later unfold into leafy shoots and flowers. Where does the sycamore keep these buds? Where are they on the other trees? How are these buds protected on the willow? On the cottonwood and the alder? On the fruit trees in your orchards? If the coverings are not distinct now, you can see them better as the buds begin to swell. Perhaps the "pussies" on the willows are already bursting out. Find out for yourselves what the willow flowers are like and where they keep their honey. The bees know, and you can watch them, they will not mind you in the least. What else do they get besides honey? Farther on we shall have a lesson on the awakening of the trees, but they do not all awaken at once, and unless you keep your eyes open all along, you will miss many of their clever ways.

CHAPTER VI.

FERNS AND THEIR RELATIVES.

How they love moisture, the ferns and their kindred! You noticed how quickly they responded to the invitation of the rain. The rock fern or Polypodium came above ground so quickly that it was hard to catch the leaves unrolling; and that is a pity, because they are such neat little balls when they first break through the soil. It is easy to see why the Polypodium can grow so fast, for the underground stem is an ample storehouse, and there is a tangle of long slender roots to gather moisture. With the first rains, moisture is absorbed to dissolve the stored food and make it ready for the use of the baby leaves that are snugly curled up like little knobs along the underground stems. Soon the leaves are above ground, uncoiling, and spreading out millions of green cells to the light. Hold up a leaf and look through it for woody strands. You can trace these strands running up the slender roots, through the underground stem, up through the leaf stem, branching through every leaflet and sending off slender fibres that reach to the very tips of the teeth along the leaf margin. As in the seedlings, these woody strands serve to carry the raw material taken in by the root-hairs, up to the green cells in the leaves, there to be manufactured into useful food.

This fern works rapidly while there is plenty of moisture, and early in the winter you will find, on the under side of some leaves, what look like tiny seeds in neat round clusters. But did you ever know seeds to come without

Fig. 26. *Polypodium Californicum.*

flowers first? Besides, if you try to make those "fern seeds" grow, you will see nothing that looks like a fern for a long time. These facts used to puzzle the people of olden times, and have led to some curious beliefs. For instance, fern seeds were supposed to be formed in some mysterious way on midsummer nights. To find them one must go alone at midnight and must repeat certain magic words. For hundreds of years no one seems to have taken the trouble to find out certainly whether the brown specks really were fern seeds. It was not until about fifty years ago that the truth about them was found out. Under the microscope they look like Nos. 1, 2 and 3 in the drawing at the end of the chapter. That is, each tiny speck is not a seed, nor even a spore, but a transparent case, somewhat like a watchcase, and it is filled with spores. So the round spots on the underside of the Polypodium are clusters of spore-cases. Each case has around it a ring of strong elastic tissue that has one weak place. When the ring breaks it acts like a spring and straightens itself, tearing open the case and flinging out the spores. Millions of spores simply perish, but here and there one will find a suitable place for growing. You will need sharp eyes to find the first growth from the spores, and you need look only in very shady, damp places. Little colonies are sometimes found beneath projecting clods of earth, or on steep north banks shaded by vines and brambles. The strangest part of the story is that the first growth from the fern spore does not in the least resemble a fern. It is a thin, delicate, green scale, not so large as your finger-nail; it lies flat on the ground and is fastened to it by delicate root-hairs. It is pictured in No. 4, Fig. 30, at the end of the chapter. Perhaps you will find a plant a little more advanced, like No. 5. The green scale has sent down a little root, and a slender stem is growing up bearing a minute ball that will unfold into a tiny leaf. Soon another leaf grows up, but by this time the first green scale is brown and shrivelled, because the little upright leaves have gone to work and can do without it. In a month or so the little plant will begin to look quite like a young fern; there will be a minute underground stem, a number of branching roots, and new leaves that look more and more like grown up fern leaves. But it will be several years before the plant will be fully grown and bears spores; then it may last for many years, that is, the leaves each spring will make more food than they need for themselves and their spores, and will send it down to the underground stem for the benefit of the next season's leaves. Examine the underground stem of the larger ferns you have collected, and see if you do not find traces of the leaves of other years.

The little fern growing from the spore in Fig. 30, is a golden-back fern, and one a year or two older is pictured in Fig. 27, No. 3. Everybody in California knows and loves this fern. The golden dust on the under side of the leaf, is so abundant that it leaves a beautiful golden imprint if you press a leaf against a dark dress. This powder is really a sort of wax, similar to the bloom on some leaves and fruits. Perhaps you can guess one of its uses. You know how quickly ferns wither when you pick them. This is because the leaf cells have thin walls that give off moisture readily. If these cells had thick walls the water current that carries raw material to the leaves would move too slowly, for ferns grow in shaded places and, in California, do much of their food-making during the winter. But the wax-like coating on the underside of the leaves protects the layer of cells against evaporation; so when hot, dry days come, you will see golden-back leaves curl up, leaving only the under surface exposed; if the dry weather does not last too long, the leaves will revive with moisture, but unless these ferns grow in very sheltered places, they die down to the ground before the dry season is over.

Fig. 27. SOME EARLY FRONDS.
1. Maiden hair. 2. Coffee fern. 3. Golden-back.

But the golden dust serves another purpose. The golden-back, like other ferns, has its pores, or openings, through the leaf-skin, on the underside only. These pores close when the plant is in danger of losing too much water, but much of the time they need to be open to allow the water current to pass out and to assist in breathing, so these pores must not be choked up in any way. Now if you have looked, you know that dew collects on the under side of leaves as well as on the upper, and it remains here longer because the sun cannot reach it. In the sorts of places ferns frequent, leaves are sometimes not wholly dry for months; so you see they need some device to keep their pores from being stopped up by water. The wax-like powder of the golden-back fern, acts like oil on the feathers of water-birds; it prevents the leaves from being wet; the water simply collects in little drops that roll off, and the water current and breathing are not hindered.

As golden-back ferns grow older, a brown powder, too, appears on the lower side of the leaves, first in delicate lines, later on perhaps covering the entire surface. The microscope shows that every grain of this brown powder is a case full of spores. Some California children know where to find the silver-back ferns. Often young golden-backs are mistaken for silver-backs, but the real silver-back fern is usually larger and hardier than its cousin with the gold dust. The upper surface of the leaf is thick skinned and rather sticky, and the lower side is covered with a silvery powder.

Have you ever thought what gives the maiden-hair fern, No. 1, Fig. 27, its name? Look at its stems. The Greek botanical name, Adiantum, means not wet. Find out if this is a fitting name for the fern. The silvery appearance of the leaves under water is due to a layer of air between the leaf and the water. So the maiden-hair fern, as well as the golden-back, knows how to escape being choked

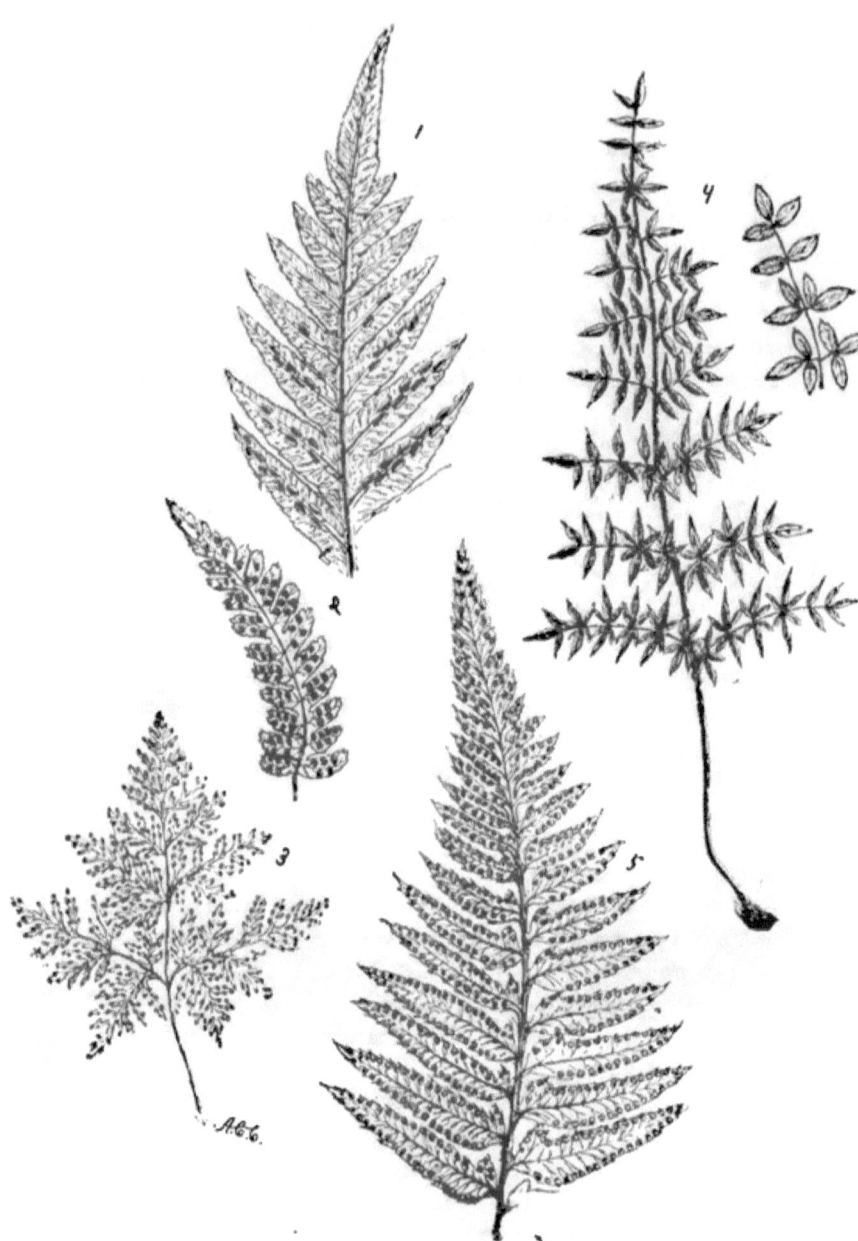

Fig. 28. FRAGMENTS OF SPORE BEARING FERN FRONDS.
1. Tip of Woodwardia. 2. Pinnule of shield fern. 3 Lace fern. 4. Bird's-foot fern. 5. Tip of sword fern.

by water; and it needs this protection, for it is often found growing in places that are dripping with moisture for weeks at a time. The leaves of the maiden-hair are very thin, that is, the green cells are spread out so as to be well exposed to what light they can get in the shaded cañons where they are found. You are likely to find the largest leaves in the dimmest cañons.

Can you think why it is better for a thin, delicate leaf like the maiden-hair to have its area broken up into little leaflets, than to form an entire leaf? Of course these fine divisions of the ferns make them very attractive to us, but they must be of some use to the ferns themselves. Remember that even in sheltered cañons, heavy rains beat on plants, and winds might tear them. You know what happens to big banana leaves in such storms. It would be very harmful, too, for fern leaves to shade one another. If your maiden-hair leaves are old enough, you can see where they keep their spore-cases. The little fragment in the picture shows this.

The coffee fern, No. 2, Fig. 27, turns back its leaf margins like a neat little hem to protect its spore-cases, and so does the bird-foot fern, No. 4, Fig. 28. You can think for yourselves what gives these ferns their names. They are the most hardy of all our common ferns. The bird-foot fern grows in rocky places that are fully exposed to the sun, yet the leaves usually survive the long dry season. In the winter you will find, growing from the same thick, fuzzy underground stem, tender new leaves that the rains have called out, and the stiff, woody leaves of the past year. The coffee fern seems to prefer more shaded places, but it is usually able to keep its leaves during the dry season. Both of these ferns, in Southern California at least, make their most rapid growth during the winter months, when their food-making might be hindered by too much moisture. But the rolled back leaflets prevent the water from clogging up the pores on the underside. So these rolled back leaf margins serve at least three purposes; they protect the spores, lessen the evaporation from the leaves during the dry season, and prevent the pores from being choked up during the rainy season.

In the drier parts of Southern California, and in mountains in different parts of the state, there are found several kinds of exquisite little ferns with leaves very finely divided. They have various names, such as lace fern, lip fern, woolly-back fern, and scaly fern; No. 3, Fig. 28, is one leaf of a lace fern. Generally these ferns have a thick skin or a waxen coat, or else they are densely covered on the under-side with tiny, overlapping, woolly or papery scales, and besides this, some of them have the habit of curling up quite snugly when the weather is too cold or too hot, and uncurling again when better times come.

Large, coarse ferns that grow in masses, are often called bracken. There is one kind that is common in California as well as in many other parts of the world. In Southern California you will find it only in the mountains among the pines, or in cañons; but in the regions of greater moisture, it often forms thickets from four to six feet high, and acres in extent. This fern seems to rejoice in heat as well as in moisture; it attains its full growth in midsummer and dies down with the frost. In some countries the leaves of this fern are used for thatching roofs, and it is said that young leaves and underground stems are sometimes cooked and eaten.

Perhaps the most beautiful of all our California ferns, is the Woodwardia, which grows in graceful clumps along mountain streams. The great feathery leaves, sometimes six or seven feet long, have a tropical look, but they are really very hardy. In Southern California they survive summer drought and winter frost, and cut leaves can be kept in the house, fresh and beautiful, for weeks. No. 1,

Fig. 28, is a fragment of one of the leaves. No. 2, in the same cut, is a small fragment of another of the larger ferns. It is called a shield fern because the coverings for the groups of spore-cases look like tiny shields. No. 5 is the tip of the leaf of another kind of shield fern. This fern has beautiful, dark green, glossy leaves that are so slender and stiff that the plant is sometimes called sword fern.

If you have searched in moist places for the first stages of ferns, you have probably found groups of other little, flat, green bodies fastened to the soil by root-hairs, but thicker than the first fern plants. As the season advances, some of these tiny plants send up, not little fern leaves, but tiny stalks that resemble toadstools or umbrellas, like No. 9, Fig. 30. Beneath each green umbrella are several little sacs of spores, which are very interesting under the microscope. These plants are called liverworts, and although they are so tiny, they know very well how to take care of themselves. Like the ferns, they must have much moisture in order to thrive. When dry days come they curl up so tightly that they seem merely dark lines on the soil; but, wet the soil, and in an hour or so the little liverworts are all uncurled, and are quite fresh and ready for work again. There is another kind of liverwort, No. 10, very common in green-houses. It crowds in everywhere, sometimes even covering the sides of flower-pots, and it is not strange that it spreads so fast, for each full grown plant has a little pocket or two full of green particles, every one of which can grow into a new plant.

The true mosses, which form a beautiful, bright green covering for damp soil and rocks, or even for walls and shaded roofs, are cousins of the liverworts. Many of the California mosses are very small, but you can readily see that each tiny plant has root, stem and leaves, and you will often find the pretty little urns in which they keep their spores. Perhaps you know some moist shaded place where

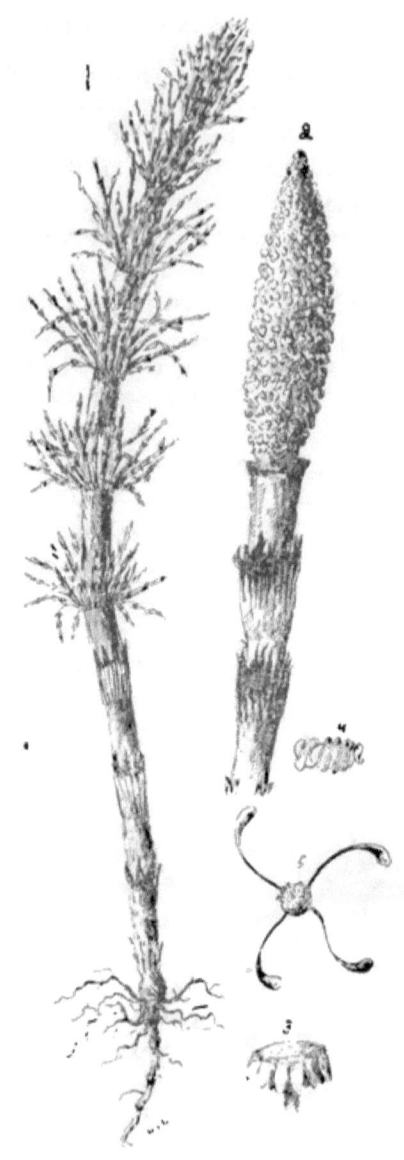

Fig. 29. SCOURING RUSH, OR HORSETAIL.—*Equisetum*.
1. Sterile shoot. 2. Spore-bearing shoot. 3. Spore-bearing branch, x 5. 4. Moist spore, x 150. 5. Dry spore, x 150.

the mosses grow large and feathery or fern-like. In countries where the weather is always cool and moist, the mosses are perhaps the loveliest of all their plants.

Should you guess that the plant in Fig. 29, the scouring rush or horse-tail, is also a relative of the ferns? It is even more particular than the fern about being in a very moist place. Very early in the spring you may find branches like No. 2, coming up from a sturdy underground stem. Later on, stalks like No. 1 develop, and grow to be several feet high before the summer is over. They are harsh and brittle and can be used for scouring. Think out the use of this hard substance to the plants themselves. The stalks that come up first end in pretty cones several inches long. You have only to handle these cones to see that they are covered with tiny branches that bear cases full of green spores. No. 3 is a single branch enlarged. If you moisten some spores and watch them dry under the microscope, they will seem to be jumping about in a very lively way. This is because each spore has four arms that coil about it when it is moist, but spring back suddenly as the spore dries. When these spores chance to alight in the right sort of place, they grow into little flat, green bodies, similar to liverworts or the first stage of the fern; but the growth from the spore is very slow, so it is best for the old plants to live on; and they do live, probably for centuries; that is, the same underground stems go on sending up spore-bearing and food-making stems year after year. In the course of time the interlacing and matted underground parts form thick peat bogs, and in some countries this peat is used for fuel.

Ferns and their relatives grow in greater variety, and attain much greater size, in countries that are hot and moist all the year round. On the Island of Jamaica there are places where one can find fifty kinds of ferns in as many yards. Some of these tropical ferns climb up tree trunks,

CALIFORNIA PLANTS IN THEIR HOMES

others become themselves small trees. Geologists tell us that ages and ages ago the earth must have been very densely covered with ferns and kindred plants, much larger than those we know to-day. Our principal coal fields have been formed by these plants, so we may sit by our firesides and enjoy the sunbeams imprisoned by ferns and their relatives countless years ago.

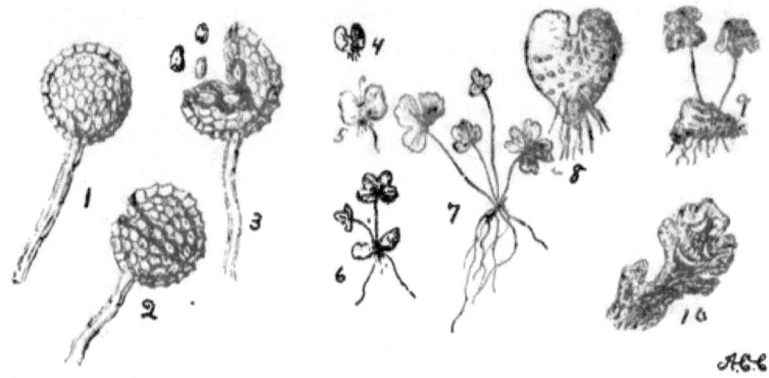

Fig. 30.

1, 2, 3. Spore-cases of ferns, x 75.   4, 5, 6, 7. Early stages of maiden-hair ferns.
8. Fern prothallium, magnified and diagramatic.   9, 10. Liverworts.

CHAPTER VII.

SOME EARLY FLOWERS.

Of our many early wild flowers, which shall we choose? A large flower that is easy to understand, is best to begin with, so we will take the peony. It is one of the earliest, too; in some parts of California its large red flowers open before Christmas time.

It has an underground stem and long, thick roots that have stored the food to give it such an early start. The peony has many large leaves, which stretch out to the sun, let the light through their pretty slashes, roll the moisture off their smooth surfaces, and behave generally like leaves that must be in haste to finish their work before the rainy season is over. They are soon able to replace the food they have used from the storehouse, but they have other work to do. There must be new peony plants; that is, there are seeds to be made, and this is the reason for the flowers.

In the centre of the peony are the three cases that will some day hold the seeds. These cases are called pistils. The beginnings of the seeds are already there, little white bodies, called ovules. The name is made from a Latin word that means egg, for each ovule contains what is called an egg cell. The little plant that a seed always contains has not yet begun to form. The part of the pistil that encloses the ovules is called the ovary. Above the ovary, at the very top of the pistil, is a pair of thin lips that have their inner surface sticky. This sticky surface is called the

Fig. 31. PEONY—*Pæonia Californica.*
1 Pistil. 2 Stamen. 3. Modified stamen. 4. Petal. 5 and 6. Sepals.

stigma. Surrounding the pistils of the peony are many bodies like No. 2, Fig. 31. Each one is called a stamen, the stem part of it is called the filament, and the upper part the anther; through two openings of the anther escapes a yellow dust, which is called pollen; and very precious dust it is, for it is only when some of it reaches the stigma and there grows down till it unites with an egg cell of an ovule, that the little plant can begin to grow; that is, ovules can not become seeds without the help of pollen. Now pollen is easily injured, both by cold and by moisture. Let us see what protects it. Around the stamens are two circles of little leaves, the inner one red, the outer, green or purple and green. The inner circle is called the corolla and each leaf a petal, the outer circle is called the calyx and each leaf a sepal. In the bud, the calyx and corolla are closely wrapped about the stamens; as the bud opens, they still protect the pollen, because the flower faces downward and the calyx and corolla are like a roof over the stamens.

But why should the bud have opened at all, if it is only necessary that the pollen fall on the stigma? Could not this happen even better in the bud? We enjoy the open flowers with their pretty colors, but of what use to the plant is their opening? Some great naturalists have tried to work out the answers to these questions, and this is what they seem to have proved. Nature prefers that flowers should not work alone at seed-making, for the best seeds result when flowers help each other, when each gets the pollen for seed-making from some other. In fact many flowers are so arranged that their own pollen cannot reach the stigmas. The peony keeps its stigmas quite beyond its anthers, and the pollen cannot easily fall on them. As soon as a peony bud begins to open, the stigmas, all ready for pollen, stand on guard at the entrance. What is to bring pollen to them? If you have fresh flowers, you will find

Fig. 32. BUTTERCUP—*Ranunculus Californicus.*

that among the stamens are yellow, cushion-like bodies, covered with honey. Now bees and other insects have found out about this honey, and if they have been to older flowers for it, they are covered with pollen. When once bees begin to visit peonies, they visit every peony in sight before calling on other flowers, and when they discover this opening bud and enter it, they must strike the stigmas and rub off some of the pollen on them; so our peony gets pollen for seed-making from some other flower. The botanists call this, cross pollination. You can see now why such flowers should not be green; the color helps the insect to find the flower. The odor, too, is an advertisement.

The buttercup is first cousin to the peony, though you may not see any family resemblance; we will talk of that in some other chapter. It has not such ample storehouses as the peony, but there is some food stored in its short underground stem and its little clustered roots, which survive the dry season. Its leaves resemble the peony leaves, and behave in much the same way, but its flowers, instead of hanging down their heads, look straight at the sun and follow it all day long. As the flowers grow older the green sepals, which in the bud help in food-making, turn yellow, bend back, and finally fall off entirely; they really are not needed, because there are so many bright varnished petals to glisten in the sun. If you were a bee flying toward this flower, you would see something else glistening,—a tiny drop of honey at the base of every petal.

Now study the flowers to see how the bees repay them for their hospitality. A buttercup has many little pistils, each of which has an ovary, containing just one ovule, and a wee, fuzzy stigma at the very top. There are also many stamens. Find a flower that has just opened for the first time. The stigmas will be ready for pollen, and the stamens will be standing back in a compact ring, not yet shedding pollen. Next examine an older flower. The inner stamens have risen up and covered the stigmas, their pollen still unshed, but the outer stamens are shedding pollen directly away from the stigmas. Keep this flower in water and examine it from day to day. It will behave very nearly as it does out of doors, closing at night to protect the pollen. You will see that the buttercup takes nearly a week to shed its pollen, the inner anthers rising and taking the place of the outer ones when they are empty. Now during all this time a bee or any other large insect visiting the flower for honey, is sure to carry away some pollen, not because it is sure to alight in the center of the flower, and so must reach over the anthers to get the honey. You may often see a brisk little bee double himself over the ring of stamens and make a complete circuit of them as he sucks up every drop of honey. He is a very dusty bee when he leaves the flower, and if he goes next to a newly opened flower, many of its stigmas will get the grains of pollen they need for seed-making.

So the buttercup, with its honest, round, shining face, is well paid for its generous hospitality. But the buttercup knows nothing about economy or exclusiveness. Its bill of fare is well advertised, and all may come; any insect, large or small, can get its honey and the pollen that collects in its petals, and little beetles, small flies and other tiny, gauzy creatures often alight on the petals, take pollen or honey, and go away without touching anthers or stigma.

There are some flowers, however, that know how to exclude such useless guests; the cluster lily, for instance. It may trouble you to make out the calyx and corolla of the cluster lily. Really there are three sepals and three petals, but they are all united to form one purple cup. The six stamens, as you can see by laying the flower open, grow on the inner surface of the cup, three of them having longer anthers than the other three. Back of these long anthers

Fig. 33. CLUSTER LILY—*Brodiæa capitata*.

are pairs of white, pointed "appendages," the botanists would say. These white teeth fit closely together and nearly close the flower; they, of course, protect the pollen, and the cluster lily needs neither to hang its flowers down nor to close them at night. But the appendages have other uses. Down at the very base of the cup is a little ring of honey; these white points form a conical cover for the cup, leaving just a bit of an entrance at the top. So only insects with long slender tongues or wee insects, called thrips, can get the cluster lily's honey.

The pistil of the cluster lily differs from those we have noticed before. The ovary really consists of three little rooms, or cells, as they are called. Above the ovary is a white stem called the style, and at its top are three little lines of fuzz; these are the stigmas, and they are quite ready for pollen when the bud begins to open. So, like the peony stigmas, they guard the entrance to the honey in the bud, and are sure to be struck by the bee's or butterfly's tongue. Now, if the guest has first visited an older flower, he must have touched the anthers, because they are close to the narrow opening left by the appendages; so every guest that comes from an older to a younger flower effects cross pollination.

Examining the stamens carefully, you will see that the three long anthers open at the side, and so shed pollen for the benefit of other flowers, but that the three short anthers discharge their pollen directly on the stigmas. The cluster lily, then, seems not to trust its guests entirely, but after a little, pollinates itself. This lily is rather ungenerous as well as exclusive, for it serves only a little honey at the bottom of its deep, covered cups. It seems to invite large bees; the cup is not too deep for their tongues, and blue is supposed to be their favorite color; but the practical bees are quick to learn that there is little honey, and they rarely visit the lilies when more generous flowers are near. Butterflies are more dainty, and come more frequently; but it is well that the flowers can pollinate themselves.

The rest of the cluster lily plant, also, is interesting. You remember how the first leaf broke through the soil after the first heavy rain; if you kept on watching the lilies, you know that the leaves grew rapidly, standing straight up to get the full sunlight; by digging now, you can find how much food they have sent down to be stored for next year's plants; they have made a new bulb on top of the old one, and perhaps have had food enough to send out a little colony of side bulbs. By this time, the work of the leaves is nearly over, and they are dying back, for they have no devices for meeting dry weather.

But the early flowers we love best of all, are the violets and shooting stars. They, too, have underground storehouses, and this is fortunate for the children of the future, because we of to-day pick so many early wild flowers that we give them little chance to mature seeds. The violet, or yellow pansy, as it is often called, is, like the cluster lily, an exclusive flower. It keeps its honey, in a cup, or spur, at the base of the flower, and the yellow petals have many purple lines pointing the way to it—honey guides, they are called. The petals are not all alike; the lower one is broad, forming a platform for the guest to stand upon when he comes for honey. The violet, too, invites only guests with slender tongues. You can see the little green stigma close against the lower petal, just where it must be struck by a tongue in search of honey. To find out about the pollen, you must tear away the petals; the stamens have no filaments, and the anthers are close together, forming a little box about the ovary. The tops of the anthers and the petals quite enclose this box and protect the pollen well.

You may have to watch a long time out of doors to see the violet visited. Really, her honey is so difficult to

Fig. 34. VIOLET—*Viola pedunculata.*

get, that her guests are few. But bees sometimes come; a bee alights on the platform petal and thrusts in his tongue; but he seems unable to get honey in this way, so he whirls about and clings, head downward, to the upper petals while he sucks up the honey. In this process he must rattle out much pollen on his tongue. He does not usually strike the stigma as he leaves the flower; but takes the pollen to another flower, and so cross pollinates it.

The shooting star has the gayest dress of all our early flowers. There are bands of crimson, purple and gold on the delicate rose or lilac petals, which shade into white. Should you not expect such an elegant flower to provide a bountiful feast? Really, fine-lady shooting star is not in the least hospitable. She serves no honey at all, and a guest that would come for pollen has actually no place to stand, but must hang, back downward, from the gorgeous stamens. Yet she expects guests to carry pollen, for the bit of a stigma is quite beyond the anthers, and at first faces downward. She takes good care of her pollen. The puffy, purple filaments are grown together so that the anthers are held closely together and the pollen is kept in until something disturbs the flower. Besides this, the flower hangs downward and the petals form a roof. You may watch for a long time on a hillside covered with shooting stars before you see a guest, but bees do come sometimes. Of course, they strike the stigma first, and then get a fresh supply of pollen on the under side of the body. Think whether this is cross pollination. But the shooting star finds it necessary to look out for herself later on. As the flower gets older, the anthers shrink, and, at the slightest jar, the pollen comes out in little clouds; strike one and see. At the same time the style turns upward and the stigma at the tip is so placed that the pollen falls on it; in this way the flower is self pollinated.

The cluster lily and shooting star keep their flowers in

Fig. 35. SHOOTING STAR—*Dodecatheon Clevelandi.*

clusters, and this plan has many advantages. One stalk will raise them up where they can be seen by insects; they are more conspicuous in the mass than if they were alone; bees can visit them more quickly, and a cluster lasts much longer than a single flower; the shooting star clusters last several weeks; the cluster lily three or four months; while a single buttercup, violet, or peony keeps its petals only a few days. As the seeds of all these wild flowers ripen, the plant must have some device to scatter them. Two of the plants shake out their seeds, two fling them, and the other has tiny hooks that help. Find out all about them for yourselves.

In collecting these wild flowers you have probably already found several other sorts. In moist, shady places you would be almost sure to find the little chickweed and the miner's lettuce. This chickweed is one of several kinds that are world-wide weeds. The miner's lettuce was used for salad by the miners in early days. Some of the other early flowers will be referred to later on, others you will have to study by yourselves.

In your gardens, too, are early flowers that you can easily watch and study by yourselves, the calla, for instance. What a fine storehouse it has! We usually give it no time to rest in our gardens because we water it all the year round, but if it is kept dry in the summer, it will take a rest, and begin life again very vigorously in the winter. Be sure to notice the little side shoots from the main storehouse stem. Each one can become a separate plant, and, as they break off very readily, it is not easy to rid a garden of callas after they have once become established there.

Notice the leaves and leaf-stems of the calla; try to imagine where the water falling on them in a rain would run off; watch and see. The grooves or channels carry it toward the centre, and it reaches the ground just where it will most benefit the underground stem with its short roots.

Fig. 36. CALLA.

The leaves of the other early plants we have been studying do this in a less degree. Contrast this with what becomes of the rainfall on an oak, or some similar tree. You often seek shelter under a tree during a rain, because little of the water reaches the ground directly beneath the tree, but at the circumference what a drenching you would get. If you have ever seen a tree uprooted, you know that the fine roots that take in water are in just this region.

But let us return to the calla and examine what is commonly called the flower or the lily. The yellow column in the centre you can break up into many little bodies that give out white powder from their tops; these must be anthers. At the base are many little bodies that are clearly pistils. Now the botanists call each pistil of the calla a flower, and each stamen a flower, so the yellow column is really a flower cluster, and the big white wrapping, which you have called the flower, protects the flower cluster and serves another purpose beside. When this white wrapping begins to unroll, you will find the stigmas down at the bottom ready for pollen, but the anthers will not shed pollen for some time. Now the bottom of the newly opened calla is a very cosy place for little insects to rest or spend the night; it affords some shelter, and is really much warmer than the outside world. If insects that have been getting pollen from older callas come to this opening one for shelter, they will cross pollinate it. Look for insects in your calla.

The Chinese lily is easily watched, and so is the iris, or flag, that will be in flower a little later.

CHAPTER VIII.

THE AWAKENING OF THE TREES.

The willows awaken first; in fact some willows hardly seem to nap at all. In autumn we saw them beginning to rid themselves of their leaves, but before the old leaves were quite gone, the brown buds along the stems were showing signs of life. Before Christmas, perhaps, some of the buds had thrown off these outside wraps, and had appeared in their inner coats of silky gray fur. Then we said the pussy willows were coming out. Some of these pussy-like buds were clearly baby leaves curled up together; others were what even the botanists call catkins, which means little cats. As the catkins grew longer, it was easy to see that the gray coat was, like an old-time armour, made of overlapping scales, and indeed it was an armour against Jack Frost and other foes of the tender plants. Soon some catkins that grew in sunny places became yellow and fluffy and were covered with golden dust. Perhaps you can find some such catkins still, and can see that they are fluffy because stamens have pushed out from beneath every little scale; the yellow dust is the pollen they are shedding.

But there is something else beneath every little scale, as the bees know very well; it is a very tiny, green peg covered with honey. Out of doors on a fine day, you can see how these minute drops of honey glisten in the sun,

Fig. 37. WILLOW.
1. Staminate flowers. 2. Pistillate flowers.

and throngs of bees come to this feast the willow provides. Often the humming of the bees helps one to find the tree. You can watch the bees as closely as you like; they are too busy to mind you in the least, for they are really doing two things at once; they are circling round and round the catkins sucking up the honey, and at the same time they are brushing off the pollen that clings to their bodies, and packing it away in the pollen baskets on their legs. The bees will have many babies to feed, and the pollen is just what they need to store for them.

But you find willow trees with catkins that are not yellow. Instead of stamens, a single pistil has pushed out from beneath each scale. Each pistil has a little box for seed, the ovary, and two, tiny, moist stigmas to catch the pollen. Each of these flowers, too, has its drop of honey for the bees.

Now let us think out how the willows are rewarded for all this hospitality. The trees that have pistils are the seed-makers, but they must have pollen in order to make good seed, and the pollen is found on other trees. Perhaps the wind carries some pollen for the willows, but the wind is a wasteful servant, and the willows prefer to pay the bees to do their work. And this is how the bees help the willows. On the catkins that bear stamens the bees fill their baskets with pollen for their own use, but they also carry away very many grains that cling to their bodies. When they fly to other trees to get the honey among the pistils, the precious dust is rubbed off on the rough, moist stigmas, where each little grain can grow and help to make a seed.

And so it happens that on the seed-making trees every catkin has its many pistils well supplied with seeds. You have seen these seeds bursting from their tiny pods and floating about on their white wings. You can see how light they are and can imagine how the winds will carry

Fig. 38. YOUNG SHOOTS.
1. Alder. 2. Poplar, or cottonwood. 3. Sycamore. 4. Fig. 5. Walnut.

them far away from the parent tree. In the midst of our deserts many, many miles from willow-bordered streams, artesian wells are sometimes bored, and at once, as if by magic, little willow trees begin to grow around the pools.

The sycamore buds that we found during the summer were snugly hidden within hollowed-out leaf stems. As the old leaves fell we found that the buds had other coverings of brown, varnished scales; and very good waterproof coats these are, protecting from dry winds and cold as well as from rain. The buds do not venture to discard their coats until March or April, and then each leaf is seen to be clad in thick, brown wool. There is a brown, woolly collar, too, wrapped about every leaf, but this wrapping instead of falling off, becomes bright green and helps in the food-making. Some of the buds, as they unfold, hang out little balls strung on slender stems. The balls are the flower clusters. Find out for yourselves which provide pollen, and which are the seed-makers. The sycamore has both kinds of flowers on the same tree, and trusts to the wind to carry pollen. The larger balls that have been hanging on the trees all winter, were flower clusters just a year ago, and now they are nearly ready to send their seeds out into the world. Perhaps as you handle them, they will suddenly crumble into a fluffy mass. On the trees they break up so suddenly that they almost seem to explode, and the wind carries the seeds far and wide on their downy floaters.

The bud coverings of cottonwoods, or poplars, are much varnished, and the young leaves, too, prefer varnish to woolly or silken coats. Notice how the leaves are rolled in the bud. The alder leaves are folded like fans, and they too have varnish enough to look very shining and new as they unfold in the sunlight. Both of these trees flower early, before the leaves appear, and the wind carries pollen for them.

THE AWAKENING OF THE TREES

The walnut flowers come later, and the leaves overtake them and get in the way of the pollen-carrying. The staminate flowers grow in catkins and drop pollen on their neighbors' backs in little heaps that will rise like a cloud with the first breeze. You can find baby walnuts not as large as peas. Be sure to notice that every one has a big rough stigma to catch the pollen in order to make the little plant grow within its shell. Now the little plants within the nut shells are richly provided with food, a sort of food that we like to eat ourselves, and so do the squirrels and other animals. You can think, then, why the walnut needs the bitter green rind until the inner shells are hard. The rinds turn brown finally, and are nearly the color of the ground, but the squirrels' sharp eyes can find them out, and their sharp teeth can gnaw through the shells. Still some nuts escape their foes, or are buried by them, and from these seeds new walnut trees spring up on our hillsides.

In our orchards, too, there is the spring time awakening of the trees. The apricot and peach trees are like clouds of pale rose or pink, and through the snowy pear blossoms the tender green leaves are seen. There

Fig. 39. PEACH BLOSSOMS

Fig. 40. PINE AND CYPRESS.
1. Staminate flowers of pine. 2. Young pine cone. 3. Staminate flowers of cypress. 4, 5, 6. Development of cypress cone.

THE AWAKENING OF THE TREES

is fragrance and honey, and the bees come in throngs and carry pollen from flower to flower, and from tree to tree. And so, as the petals fall, the little fruits begin to appear. Watch them as they grow. Perhaps you will discover that the apple and quince behave differently from the peach, apricot and cherry. Did you ever think why fruits should be green and unfit to eat before the seeds are ripe, or of what use to the trees are the bright colors and delicious flavors of ripe fruits?

Do you know the secret of the fig trees, that a fig is really a hollow stem with flowers inside? The fig trees common in California orchards have pistillate flowers only. You can easily see this for yourselves in young figs. Since there are no flowers with stamens to provide pollen, the pistils are not able to make good seeds, so we must get new fig trees by planting cuttings from old ones. Fig trees have been brought to California from other countries. In their home country there are many fig trees that produce pollen-bearing flowers, and the pollen is carried by very wee wasps that can crawl through the tiny openings at the end of the fig.

There are other trees, as you know, that never drop all their leaves and take a complete rest; but most of them are sluggish during the autumn and early winter, and in the early spring time they, too, are awakened to more active life. The live oak puts on a beautiful new spring dress of pale green, and hangs out long catkins that furnish pollen. If you have sharp eyes, you can find the tiny flowers that will become acorns if the wind brings the pollen.

The pine trees cautiously push out fresh green needles. They produce also clusters of pale yellow cones, filled to bursting with floury pollen. At the slightest jar, the pollen rises in yellow clouds; in pine forests the lumbermen call these clouds sulphur showers. If you look at these pollen grains under the microscope, you will understand why they rise; every one is provided with two air bladders, each as large as itself, and these serve as floaters. So, although the seed-making cones are usually high up in the trees, they get plenty of pollen. See if you can find pollen-producing and seed-making cones on the cypress or other trees in parks or gardens. Find out if other evergreen trees, the orange, Eucalyptus, pepper, acacia and so forth, have the spring time awakening; perhaps irrigation interferes with the natural habits of some of these trees.

Then there are the smaller trees, the shrubs as we call them. Some are wakened by the first rains. There are California lilacs that send out flowers in December, and other kinds that take their places later on. So for several months there are sunny slopes that are misty-blue with lilac flowers. Each tiny delicate flower spreads a feast for insect guests, but later on, puts pollen on its own stigmas.

The poison oak is one of the first shrubs to waken and several of its relatives have winter or early spring flowers. The currant-gooseberry family have very early habits. Before the lovely pink and white currant flowers and the fuchsia-like gooseberries are gone, other kinds are in flower. There are slender, yellow currant flowers that furnish honey to the earliest wild bees, and long, scarlet gooseberry flowers that entertain humming birds and the largest bees. The blackberries are quite awake in March. In some parts of the state, tree poppies flourish on mountain sides or in sandy washes.

Before the snows have disappeared from the mountains, the manzanita bursts into bloom; and such exquisite flowers as they are, those pink and white waxen bells! Their delicacy is a surprise too, for the manzanita is a very rugged plant. Its thick, red stems branch and twist and interweave so that they form an almost impassable thicket, and the sturdy leaves stand erect and brave summer heat and winter snows. The flowers provide so much honey that humming birds are glad to come with the bees. The stamens are like little pepper boxes, and the guests must strike the handles and shake out the pollen. The madrone, which is larger than the manzanita, has quite as wonderful flowers.

There are many other interesting mountain shrubs that help make up what we call chaparral. In summer time the wild mahogany has little fruits with silky, silver-gray plumes, and the greasewood fairly whitens the mountain slopes with its plume-like flower clusters. The wild cherries bloom earlier; some of them have very beautiful evergreen leaves.

Higher up in the mountains in Southern California, but nearer the valleys and coasts in the northern part of the state, are the majestic cone-bearing trees; the cypresses, redwoods, the big trees, firs and pines, trees that have few equals in the wide world. John Muir, who knows and loves California mountains so well, has written much about these trees, and they have other friends who write and speak eloquently about them, and who are making great efforts to preserve our forests. But after all, their beauty and grandeur cannot be expressed in books, and very fortunate indeed are the California boys and girls who can go to the mountains themselves, and learn to know the trees at first hand.

CHAPTER IX.

SOME SPRING FLOWERS.

Our California poppy is known and admired the world over; but under cultivation it grows pale, and it is only here, in its native home, that it can be seen in all its splendor. We have turned thousands of acres of poppies into grain fields, orchards or city lots, but we still have left in our foothills and upland valleys, glowing poppy fields that are a marvel to strangers and a never-ending delight to ourselves. The Spanish people named this flower "cup of gold;" but the botanists called it Eschscholtzia.

And why is our poppy so successful? We cannot hope to learn all of its secrets, but some of its ways are easy to understand. It has an underground part that will last for years; besides, it will come up quickly from the seed. The leaves are just the sort to make the most of a short rainy season, and the flowers take the best of care of their golden pollen; they open late and close early on fair days, and not at all in rough weather. Look into an open flower, and see how the petals hoard the pollen as it falls from the anthers. It offers no honey to guests, but the pollen is free to all who call on sunny days between 10 and 3 o'clock. The poppy takes lodgers, too, and several kinds of insects choose to sleep in this golden palace. So the poppy receives a fair share of insect attention, but not so much from bees as from flies and beetles. Some of the beetles are boorish

Fig. 41. POPPY AND CREAM-CUP—*Eschscholtzia Californica* and *Platystemon Californica*.

enough to eat the petals of the flowers as well as the pollen. But the guests must carry much pollen for the poppies. It is generally believed that, although their own pollen may fall on their stigmas, they do not mature seed unless pollen is brought from other flowers.

The cream-cups are cousins of the poppies, and have the same habit of dropping their cap of sepals as the petals unfold. This habit will help us to recognize other members of the poppy family, including the cultivated poppies. You may find the tree poppy with large, pale yellow flowers, or a pretty, little, bright red poppy whose petals fall off at a touch, or in the shady nooks of the cañons, a very delicate, little, star-like, white poppy. In sandy washes, later in the season, the prickly poppy will send out great white flowers with crumpled petals and a great many yellow stamens. In some cañons in Southern California, there is found, in May, a great, white poppy five or six inches in diameter, the plant being sometimes seven feet high. This giant poppy is called Coulter's poppy, or the Matilija poppy; it is being introduced into our gardens.

And the mustard! No amount of cultivation seems likely to drive out this common weed. In Southern California it forms thickets so high that men on horseback can be quite hidden in it. It is true here, as in Palestine, that it grows with marvelous rapidity from the least of seeds, and that the birds lodge in the branches. It matures during the rainy season; in summer time there remain only the dead, gray stalks, from which the birds gather seeds. It is impossible to explain fully why the mustard can grow so rapidly and become strong enough to drive out other plants that we take great care to preserve; but some of its advantages are easily seen. It has rough leaves, with a biting taste that most animals must dislike, and it has a great abundance of flowers, in clusters that last a long time. It provides honey, and is visited by bees that effect both close

SOME SPRING FLOWERS

and cross pollination. Finally, each plant matures thousands of seeds that can retain their vitality for years. These are scattered far and wide.

You have noticed that mustard flowers have four sepals, four petals, six stamens and one pistil. There are several other very common little flowers that have their parts in exactly the same numbers; see how many you can find before we take up the chapter on plant families.

There is a pretty spring flower, Fig. 42, commonly known as the primrose, which you might take to belong to the mustard family. It has four sepals and four petals, but eight stamens, and its pistil is a puzzle; it is easy to find the little ball-like stigma and the slender style, but no ovules are to be found within the flower; they seem to be in the stem, instead. Now what holds the ovules must be the ovary, so the apparent stem is really the ovary, and it is called an inferior ovary, because it is below the rest of the flower. This primrose appears rather early in the spring; at first there are a few pale yellow flowers close to the ground, in a rosette of leaves; later on, numerous branches spread out flat on the

Fig. 42. PRIMROSE—*Œnothera bistorta*.

ground, bearing many flowers and queer, twisted pods. The primrose flower has usually a brown spot at the base of each petal; it produces a little honey, and the brown spots help the insects to find it, but the flower is not very hospitable, and can pollinate itself.

Really, this flower is not a true primrose; it belongs to the evening-primrose family, and its botanical name is *Œnothera bistorta*. There is an Œnothera you are pretty sure to find on the beaches; its full name is *Œnothera cheiranthifolia*, variety *suffruticosa*, but notwithstanding its ugly name, it is a pretty plant, with its silken, silvery leaves and pale yellow flowers. It is a sturdy plant, too; in spite of all the intense light and heat from the sand, it keeps its leaves and goes on flowering all the year round. Can you think why this is possible? You are almost sure to meet other handsome members of this family later on, and you can always recognize them because of the inferior ovary and the parts in fours.

Now, if possible, make a collection of the following flowers:—Gilias, baby-blue-eyes, Phacelias, forget-me-nots, both white and yellow, nightshade, and morning-glories. The motto of this group of flowers might be, "In union there is strength," for all of them have their petals united into tubes, cups, funnels, wheels or something of the sort. All of them, too, have the stamens growing on the corolla. Notice the number of parts; five always, until you come to the pistil. Now, many of these plants are annuals, that is, they live but part of a year; so it is very necessary that they make good seed, and all of them invite insects to carry pollen.

The Gilias have always one ovary, which is three-celled, and one style, but three stigmas. There are more than seventy kinds of Gilias in California. The slender one in the picture, No. 2, Fig. 43, is *Gilia multicaulis*, which means a Gilia with many stems. Its flowers are blue,

Fig. 43. GILIAS.
1. Ground pink, *G. dianthoides*.   2. *G. multicaulis*.
3. Mountain pink, *G. Californica*.

sometimes very pale blue, or nearly white; they are rather small, and grow in small clusters, but they are very fragrant and furnish a goodly supply of honey. No. 1 in the picture is *Gilia dianthoides*. It is the well-known ground-pink of Southern California, and it sometimes actually carpets the ground. It is an exquisite little flower, with its lilac or pink, satin petals, slashed and fringed at the edges, and banded with crimson, yellow, and brown at the base. But this elegant Gilia is not generous in supplying honey, so, while butterflies, who seem to prefer finery to food, usually choose fine-lady dianthoides, the less showy blue Gilia is the bees' favorite. On a whole hillside, gay with wild flowers of many sorts, you will often find the bees selecting only this modest blue Gilia.

Both Gilias keep open house on pleasant days, but only from about 10 in the morning until 3 or 4 in the afternoon. They open their anthers early, making them into little pollen-covered balls, which stand guard in a ring about the entrance to the honey. Usually they hold their stigmas above the anthers, and do not unfold their own stigmas until after they have furnished pollen for those of other flowers.

The stigmas of the little blue flower often lie against the lower edge, and you can distinctly see the bee strike them as he thrusts his head into the flower; you can also see that his head is dusty with the blue pollen of other flowers. *Gilia dianthoides*, after a while, curls down its stigmas among its own anthers, and flowers of both Gilias are likely to pollinate themselves as they close at night. Besides this, the fallen pollen collects all down the corolla tube, and as the corolla finally falls off, some pollen is sure to be brushed against the stigmas; so these Gilias can pollinate themselves if insects fail them.

The other Gilia in the picture is *Gilia Californica*, or the mountain-pink. The plants are shrubby, and are two or three feet high. The flowers are very handsome and showy, but it is really one of the most unamiable of the Gilias; if you have ever tried to pluck it, you know how savage the leaves are; and the great, lovely, pink flowers seem to provide no honey at all; besides this, they keep their pollen away down in the narrow corolla tube, apparently only for their own use, for it falls directly on the stigmas. I have never seen this inhospitable flower visited, and its attractive corolla seems to be of no use to the plant.

The botanical name for baby-blue-eyes is Nemophila. There are several kinds quite common in the state. In Southern California the one most common in the cañons has very delicate blue flowers, while the kind that grows in open places has larger, deep blue flowers. But the most clever Nemophila is the one in the picture, Fig. 44; it is common in very shady places; it has not blue flowers at all, but large, dull violet ones. It is rarely gathered, because the stems are so weak that they break in the handling, and the whole plant is very prickly. A bit of its epidermis under the microscope is shown in the picture; the prickles, or

Fig. 44.
CLIMBING NEMOPHILA,
Nemophila aurita.

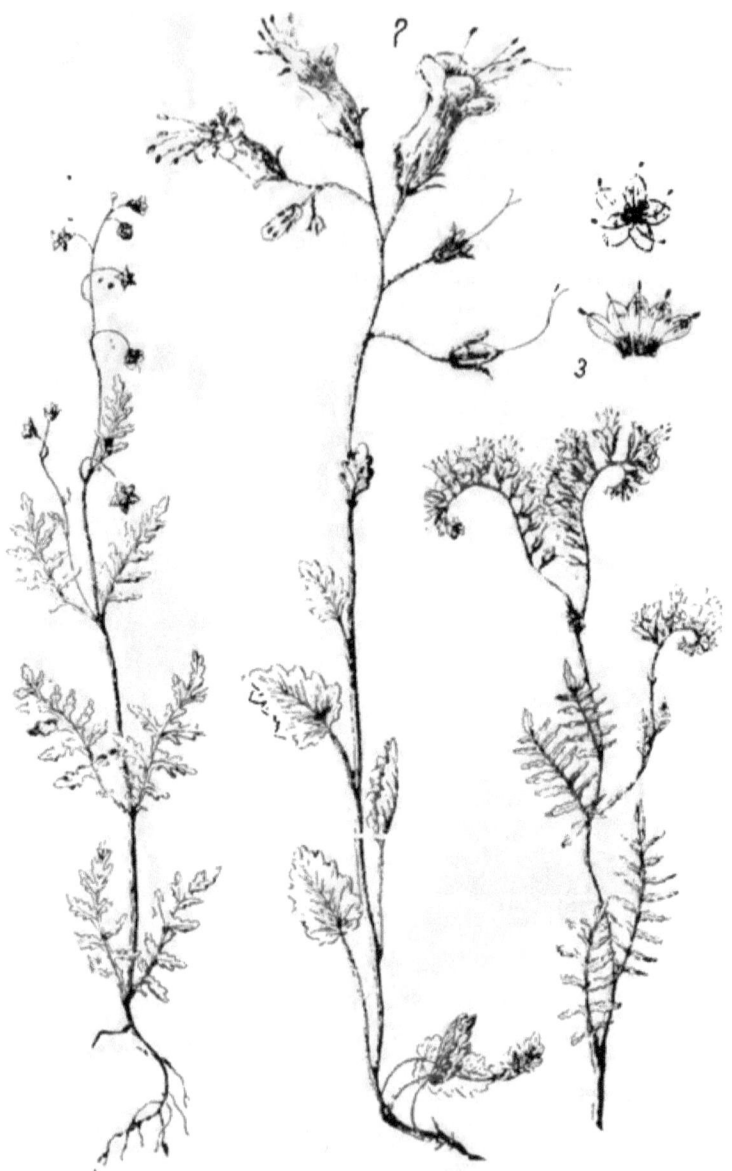

Fig. 45. SOME MEMBERS OF THE BLUE-EYES FAMILY.
1. *Ellisia chrysanthemfolia.* 2. *Pracelia Whitlavia.*
3. Wild heliotrope, *Phacelia anacetifolia.*

more truly hooks, serve at least two purposes; they defend the plant against enemies and help it in climbing. This Nemophila has very little woody tissue, but by hooking itself, by leaf, stem, or calyx, to any convenient support, it gets up into light and air, and is able to display its flowers to the insect world. I have seen Nemophilas that, as early as February, had climbed to the top of a high fence, and were looking over for something else to grasp. The flowers know well how to secure insect help. At the base of each flower are ten tiny saucers that hold the honey; these saucers fit together in pairs, forming five little hollow balls; so the honey is protected from dust, wind and rain, and is reserved for large insects, since small ones could not open the balls. The anthers face upward as they shed their pollen, so guests are sure to carry some away; later on, the two stigmas are held up where they are sure to be struck.

The Phacelias belong to the same family as the Nemophilas, and a very vigorous branch they are! There are many kinds of Phacelias; the flowers vary much in size and in color and form, but they always grow in clusters that last a long time, and coil at the ends like a scorpion's tail; the botanists call this form of cluster "scorpioid." Phacelias are usually rough or sticky, and know well how to defend themselves. No. 3, Fig. 45, is *Phacelia tanacetifolia*, a common, wayside weed in some places in California. Its flowers, like those of the Nemophila, have ten scales on the corolla tube, but the scales are not at the very bottom, and instead of being saucers to hold the honey, they fit closely together, forming a floor, or false bottom, a little way above the real base of the flower. The honey is at the very base, and the plates above fit so closely that only a strong slender tongue can get it. So this Phacelia keeps its honey for the larger and more useful guests; the bees know this very well, and come in throngs, carrying pollen from younger flowers and rubbing it on the stigmas of older ones.

Fig. 48. WHITE FORGET-ME-NOT—*Plagiobotbrys notbofulvus.*

No. 2, Fig. 45, is *Phacelia Whitlavia*, with great, bell-shaped flowers of royal purple. Its flowers are really magnificent, but they are not so clever as those of the other Phacelia. You can read their story for yourselves if you watch the flowers out of doors on a still, sunny day. No. 1 is a small specimen of another of the baby-blue-eyes family that is very common in shady places. It has a pretty first name, *Ellisia;* the rest of the name is *chrysanthemifolia*, which means only chrysanthemum-like leaves. Really, they are much prettier than chrysanthemum leaves; they are often mistaken for ferns early in the season before the dainty little white flowers appear. The flowers provide a little honey, but as they are very small, and grow in shaded places, the bees are not sure to seek them, so they contentedly pollinate themselves if guests fail to come.

The forget-me-not, or heliotrope family, like the Phacelias, has always scorpioid flower clusters. Our forget-me-nots have small flowers, but perhaps you can make out that they have one style, one stigma, and an ovary that breaks up into four parts. There are many kinds of the white forget-me-nots. The one in the picture blooms early; in Southern California, in February and March, it actually whitens grassy slopes and meadows, and gives out a delicious fragrance; and the botanists have called this dainty blossom *Plagiobothrys nothofulvus!* The flowers are too small to provide much honey, and they seem able to pollinate themselves, but they are sometimes visited by small bees, flies and butterflies. They remain open at night, and, because of their whiteness and fragrance, must attract night guests also. The yellow forget-me-nots are such coarse weeds that we hardly like to call them forget-me-nots; in some places the children call them woolly breeches. They are sturdy successful plants. They grow rapidly, are well armed against all foes, and are on the best of terms with the bees. They have, too, excellent devices for scattering their seeds, as you can see for yourselves.

The flowers of the nightshade and morning-glory place them in this group of plants. Unlike most of the plants we have been considering, these are usually perennials. Some species are world-wide weeds. As we noticed in the

Fig. 47. NIGHTSHADE—*Solanum* **Douglasii**.

autumn, the nightshade can flower all the year round in our climate; and one European morning-glory seems able to drive out all other vegetation in some California fields. Watch the climbing of our native morning-glories and compare with the chilicothe, the poison oak and the *Nemophila aurita*.

The most attractive nightshade of Southern California is found in the foot-hills; it has large and very fragrant blue flowers. Unlike the flowers we have been studying in this group, the nightshade provides no honey, so the corolla needs no tubular part. Its five, large, yellow anthers lie close together, forming a cone that reminds us of the shooting star anthers. Disturb this cone slightly, and a little cloud of pollen rises. It is shed through minute chinks at the ends of the anthers; indeed, we might call these pepper-box anthers. The bees seem to know all about this, and you will sometimes find numbers of them collecting pollen from the anthers, always striking the stigmas first. The more common nightshade, pictured in Fig. 47, has smaller, white or bluish-white flowers; it is not so attractive to bees as the fragrant, blue-flowered one. The nightshades have the same device for self pollination that we found in the shooting star; the style of the older flowers turns upward, holding the stigma where pollen will fall upon it. The morning-glory keeps its honey curiously hidden where the bees seldom try to get it, but they strike the stigmas when they come for pollen. The morning-glory, too, seems able to pollinate itself.

But surely the queen among all our California spring flowers, is the Mariposa lily, or butterfly tulip. As seen in the picture, the Mariposas are of two distinct forms. There are dainty nodding lilies, sometimes called globe tulips, the white one of the picture, the satin-bell or fairy's lantern, and a yellow one called golden lily-bell, are the most common. Their petals open very slightly, and are covered with

Fig. 48. MARIPOSAS.
1. *Calochortus albus.*   2. *Calochortus Catalinæ.*

long hairs within ; so that the children call the flowers cat's ears. They serve some honey, and entertain some guests, but they are able to pollinate themselves. There are also the larger, upright, cup-shaped Mariposas that are much more common, and few flowers can rival these stately chalices in beauty of form or color. In the valleys the lilies are usually lilac, cream or pale rose, with crimson spots, or yellow with golden brown or purple markings ; a desert species is flame color with royal purple honey guides; in the mountains there are azure, violet, purple and intermediate shades with spots and rings of other tints. And of what use is all this beauty to the plant itself? Really, the Mariposa does not calculate closely in her hospitality, as we shall see. Honey is served in the tiny bowls at the base of each petal, and the protecting fringes and borders, in elegantly contrasting colors, serve also as honey guides. This honey is accessible to many guests; indeed, there are spiders that find it worth while to assume the exact color of the flowers, and to lie in wait for the Mariposa's insect guests. The Mariposa serves pollen, too, opening anthers so slowly that the supply lasts for several days. She affords shelter also ; there is a fuzzy, homeless bachelor bee that chooses to spend his nights in this stately palace; you may find him early in the afternoon, already snuggled down for the night, standing on his head, his antennæ tucked neatly back. Now it is not until the petals begin to fade that the Mariposa's own stigmas are exposed, and even then an entering guest is not sure to strike them ; so, many of the visits paid to flowers must be of no use to them. Let us hope that the bachelor bee brings pollen, or that in nestling down he scatters some of the stored pollen on the stigmas. The Mariposa has bulbs as well as seeds, and however greedily we may pluck the flowers, we usually leave the bulbs ; so possibly our Mariposas may be spared to us for many years to come.

CHAPTER X.

PLANTS WITH MECHANICAL GENIUS.

In olden times if a man's father were a baker, he himself would be a baker, and so would his son and his son's son. Every man followed the trade practiced for generations by other members of his family, and so all became skilled workmen. This is true in some countries to-day. It is also true that there are some families of musicians or even of literary men.

In the plant world, we have one family at least, the pea family, with a talent in one direction; nearly every member of this family has some ingenious mechanical device. It is a large family and furnishes us many beautiful and useful plants. It includes the lupines, which make beautiful so many spots in California, from the sea beaches to the very mountain tops. The lupines adapt themselves to all conditions. There are annuals that grow rapidly, and flower and fruit during the few months of the rainy season; and there are perennials that, even in Southern California, can keep on blooming all the year. On sea beaches, the lupines send out roots sometimes thirty feet long, and clothe themselves in woolly or thick, silken coats; along streams, the smooth, bright green leaves of one lupine are six or eight inches across, and the flower clusters reach up higher than a man's head; while up in the mountains there are little perennial lupines but a few inches high, with leaves soft and silky as seal skin, and sturdy little stems and roots that

Fig. 49. LUPINE—*Lupinus sparsiflorus*.

store food. Some lupines come up year after year, like weeds, in cultivated land; and several kinds have leaves that know how to fold at night, or during dry winds.

The flowers of all lupines grow in clusters that last for weeks, sometimes for months, and they all have the same clever device for pollination. Take the cluster of any large-flowered lupine you can find, and examine the flowers carefully. The calyx seems to consist of two parts; really, there are five sepals united in two groups. There is one very large petal that stands upright and is called the banner; the other four petals form a platform for guests, and enclose the stamens and pistil; the two outside petals, which are called wings, usually cohere slightly at the tip; besides this, they are fitted very neatly into the inner petals so that they act with them; see for yourselves just how. If you have one of the younger flowers, you will find the two inner petals so much united that you are likely to mistake them for one, but notice that at the base they are quite distinct, and that there is also a tiny opening at the very tip; these partially united petals are supposed to resemble the bottom of a boat and are called the keel. Snugly tucked away in the keel are the ten stamens, and within their united filaments is the pistil, the style and stigma projecting slightly beyond. To understand all about the stamens you need to begin with the bud; you will find that the five higher, longer anthers shed their pollen in the keel before the banner rises, while the other five, by means of their thickened filaments, hold the shed pollen firmly in place in the tip of the keel.

Now, when a bee comes to call on a lupine, he is sure to alight on the lower petals; imitate with your pencil his weight on this part of the flower, and see what happens; there is always a little jet of pollen forced out. You can think how the lupine does this. The pollen has been packed away in the tip of the keel and held there by the filaments; as the weight of the bee presses down on the keel, the stiff stamens push up the pollen and force it through the opening at the tip. So the lupine lets its guests pump out pollen. The bees seem to appreciate the pollen very much, for, in spite of the fact that the lupines provide no honey, the larger and more fragrant kinds are much visited by bees. It is most entertaining to watch them stow away the pollen in their baskets as they pump it out. Hive bees can visit about twelve flowers per minute, but a great bumble-bee can pump out and pack away the pollen of thirty-five flowers in the same time. Since bees are so swift and industrious, do you wonder that many flowers favor them for guests? Perhaps this is why most of our lupines have attained the bees' favorite color, blue. Of course you can see that at every visit the bee first strikes the stigma, which is mature in the older flowers; that is, he cross pollinates the flowers.

The lupines have also a mechanical device for scattering seed, a device that is used by other members of the pea family. The fruit, you see, is a kind of a pod; it is called a legume, and since all members of this family have the fruit, a legume, the Latin name of the family is Leguminosæ. The legume of the lupine has along its edges an elastic tissue that causes the two parts, when separated, to coil and twist back with considerable force, so scattering the seeds.

The alfalfa is a member of the family Leguminosæ. You will recognise the family likeness at once when you look at the flowers, for they have banner, wings and keel, as the lupines have. The alfalfa is a European plant, introduced here by way of Mexico and Chili. It is a great boon to our western country, where the upper layers of soil become so dry during the rainless months, for the alfalfa roots will grow down a considerable distance to find a moist subsoil; we sometimes find it growing in waste

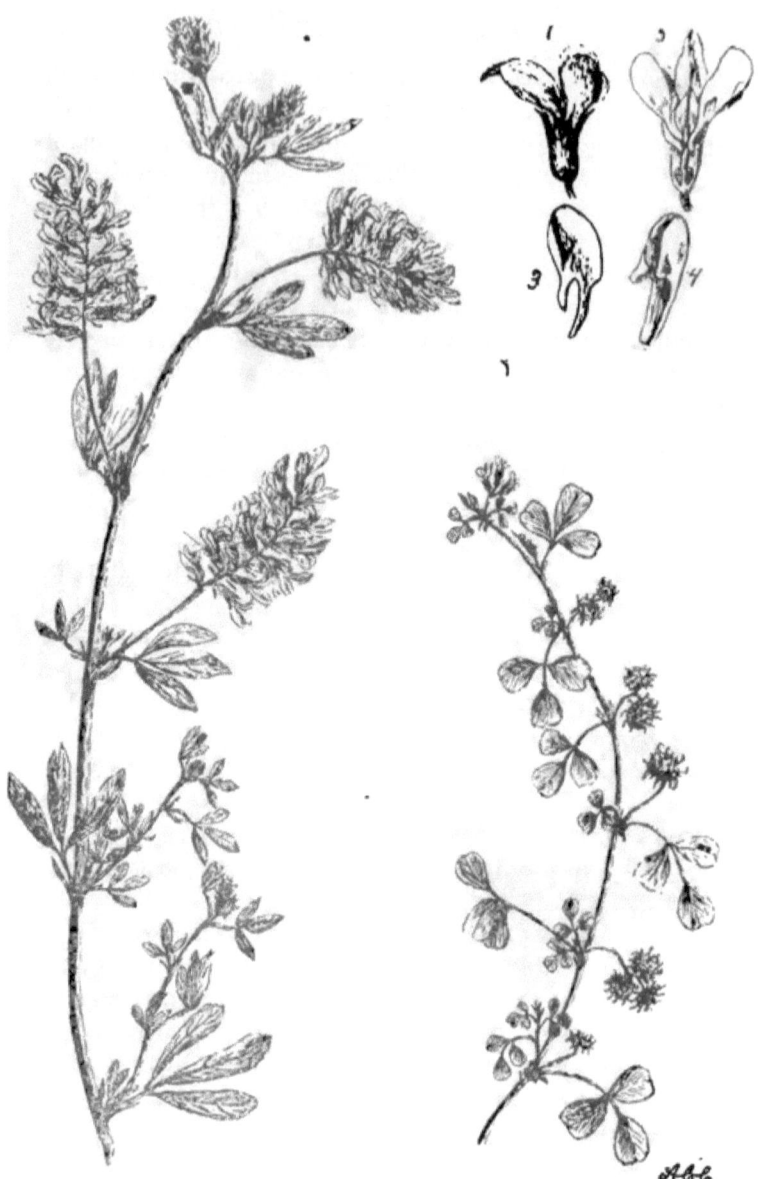

Fig. 50. ALFALFA AND BUR-CLOVER—*Medicago sativa* and *Medicago denticulata*. 1. Exploded flower of alfalfa. 2. Unexploded flower with banner removed, top view. 3. Wing. 4. Keel.

LUPINE, ALFALFA AND BUR-CLOVER

places year after year without cultivation. The flowers of the alfalfa, too, have their special mechanical trick. Examine a cluster of flowers carefully, looking each flower squarely in the face. Perhaps you will find that some flowers, like the enlarged one in the picture, have the column of stamens and pistil close against the banner, while in other flowers nothing of the sort is there. Take one of these latter flowers, and imagine where an insect seeking honey would thrust its tongue; imitate this action by inserting a needle, and instantly, the stamens and pistil fly up like a "jack-in-the-box." Now let us look for this machinery. Remove the calyx and banner, and push back the column of stamens and pistil into the keel. You will see that there are two little projections, one from each wing, that are fitted over this column, also that the wings are firmly fastened to the keel by means of other projections that fit into pockets. So the column was at first held down in the keel by means of these projections from the wings. The honey is at the very base of the stamens, and is reached only through two little openings on the upper side. When an insect alights on the lower part of the flower, and thrusts his tongue directly in for honey, as you did the needle, the tongue separates the projections, and the stamens and pistil fly up with much force and strike the tongue. The pistil is likely to strike first, and the whole tongue becomes newly dusted with pollen.

Of course this is a very rude way for the flower to treat its guest. The blow must be disagreeable, and the bees and butterflies seem to dislike it, for some of them have learned to get honey by inserting the tongue at the base and side of the flower; in this way they obtain honey without touching the stigma or pollen, and the flower, on account of its rude behavior, gets no help from the guest. Indeed, while the alfalfa has a very ingenious mechanism, it has really not so good a method as the lupine, for only the first guest strikes the alfalfa stigmas and pollen, while the lupine furnishes pollen and has stigmas struck over and over again; the pollen of the lupines is always protected, while the alfalfa anthers are quite exposed after the explosion of the flowers. The only advantage of the alfalfa is that butterflies, as well as bees, can pollinate the flowers.

Now the most wonderful part of the story is, that the wee flowers of the bur-clover have exactly the same machinery as the alfalfa. Their pistil and stamens, too, jump up like a "jack-in-the-box" and strike the guest's tongue; but the column of stamens and pistil in the bur-clover is so short that it cannot trouble the insect much, and after it has sprung up against the banner, it is still protected by the wings. This makes it difficult for you to see the flower explode, but if you have patience and good eyes, you can find the projections on the tiny wings, and the pockets in the keel. The bur-clover flowers, in spite of being so tiny, furnish a good amount of honey, and the bees seem to find it more readily than we can.

Now take a stem of bur-clover, and note the changes from the flower to the bur. Would you have thought that the bur is really the same sort of fruit as the lupine pod? By looking closely you will see that the bur is at first a tiny pod, and that it grows two rows of teeth on one edge, takes a little turn, then twists again and again, until it is the hooked coil, or bur, that fastens itself so readily to our clothes. The alfalfa, too, has a coiled pod, but it is not provided with teeth.

The true clovers are another branch of this ingenious family, Leguminosæ. We have some very pretty clovers in California, but they are not such an important part of the vegetation as they are in countries with more moisture, in our Eastern States or England, for instance. A head of clover is, of course, a flower cluster, and you can easily see that, like the lupine and alfalfa, each little flower has banner, wings and keel; that is, they are what are called "butterfly" flowers. If you have patience, you can find that the clovers, too, have mechanical contrivances for keeping wings and keel united, and that they serve honey. Some clovers trust absolutely to their insect guests, and cannot pollinate themselves at all. This is true of a beautiful, large-flowered clover that is much used as a pasture plant in many countries; its flowers are so long that they must have bumble-bees to carry pollen for them, and if we try to grow the clover where there are not enough big bumble-bees, it will not produce good seed. So they jokingly say that England owes her beautiful clover fields to the old maids that keep the cats that kill the mice that eat the bumble-bees that pollinate the clover.

Our wild peas have such large flowers that you can very easily find their honey and the beautiful devices by which the wings and keel are held together. They have, also, little brushes on their styles for sweeping out pollen, a little for each guest to carry away. Be sure to watch how the wild peas climb. There are several kinds of wild peas that you are likely to find, and also many other common kinds of Leguminosæ. There are the "rattle-pods," so common in sandy soil; this plant is called loco weed by the stockmen, and is believed to loco horses, that is, to make them crazy. Then there is the Hosackia, or Lotus, group of small Leguminous plants. One kind is pictured in Fig. 11, Chapter III, and you are likely to find others. In your fields and gardens, too, you will find many members of this family, some useful like the peas, beans and peanuts, and some with very showy flowers; see how many you can recognize, and try to discover the devices of their flowers and fruit.

There are other families with mechanical devices,

Fig. 51. FILAREE—*Erodium cicutarium*.

especially for scattering seed. The Geranium family, which includes the filaree as well as the cultivated geraniums, is an example. The filaree, like the bur-clover, is believed to have been brought to California from Europe, perhaps with the grain seed imported by the fathers in the good old mission days. The filaree and bur-clover take kindly to our climate, the rainy season giving them all the time they need to mature their numerous seeds; and how they have spread themselves over this great state of ours! from the lovely fertile valleys overlooking the sea, where the fathers loved to plant their missions, up to almost inaccessible mountain meadows two hundred miles from the coast, and thousands of feet above it; for with the filaree and bur-clover came the sheep to carry the seeds in their wool wherever sheep can go.

You all know the filaree fruits; perhaps you call them clocks; let us trace them back to the flowers. Watch the little bluish flowers on your way to school in the morning; notice how they turn to the sun, and find out where they keep their honey. The filaree, like the bur-clover, practices such bountiful hospitality as it must have seen in the old California days. Every flower provides five generous drops of honey, which glisten in the sun and are free to all who come. The pollen and stigmas are kept where the larger guests will strike them; but the smaller ones will not pay for their entertainment, and the filaree seems quite able to pollinate itself; at any rate, it always ripens an abundance of good seed. After the petals fall, the five little pistils in the centre of the flower, also the part of the flower that holds them, the receptacle, begin to grow, and they grow, and grow, until they are about two inches long. Now if you will sit down in the midst of the filaree on a warm April or May morning, you can see the little brown "clocks" shooting off from the parent plant in every direction; you can even hear the soft patter as they fall. Pick a filaree fruit that has turned brown and is just ready to explode, and give it a little twist. The five pistils spring away from the receptacle like bits of bent whalebone, and for much the same reason. As they fly off, they may be caught and carried with the wind because of the long silky hairs of the styles.

Now look carefully at the seed part, or rather at the brown ovary wall that contains the seed; it is tipped with a little hook and is covered with bristles that slope outward and upward. Put one in your sleeve and see how easily it slips in, and how troublesome it is to get out. So these bristles must be of great use to the seeds in keeping them in the sheep's wool until they are carried a long distance. The silky hairs, then, are of use when the wind is the carrier; the bristles, when the sheep serve the plant; but the most interesting feature is the twisting motion that gives the fruits their name of "clocks." The clock you have put in your sleeve has probably turned round several times while you were reading this. Let it finish twisting, then put it in water and watch it; before long it will be perfectly straight; as it dries, it twists again, and so on. Think of the use of the coiling and uncoiling to the little fruits that lie in the dust on the ground through all the dry season. With every dew or fog they uncoil; as the sun comes out, they twist again. Do you not see that this helps to bury them in the dust? If there is moisture enough to soften the soil, it makes the clocks uncoil at the same time, and they will actually screw themselves into the earth; so by the time the autumn rains come, our filaree seeds have pretty well planted themselves; we have seen how promptly they spring up after the first rain.

The geraniums of our gardens have the same mechanical devices for seed distribution. The pollination of their flowers, too, is very interesting, and you can find out their story for yourselves. Notice whether the flowers are regular; find the honey; think whether all the guests could get it; where the guest would alight; whether he would strike the pollen or stigmas in the younger flowers; in older ones. Then watch for the guests. What sort of guest do you think might choose the brilliant scarlet flowers? Which colors would be the best for the night moths?

There is still another family that uses either twisting or bending and unbending movements to scatter seed,—the grass family. Several of our California grasses have this habit; see if you can find them.

CHAPTER XI.

PLANTS OF HIGH RANK.

Men attain high rank because of marked ability; plants, too, are considered of high rank when they are specially well fitted to accomplish their ends. Of course the flowers of higher rank unite their petals to secure the more desirable guests, and many of them make their corollas two-lipped, using the two upper petals for one lip, the three lower for the other; the lower lip usually serves as a platform for the guest, as we shall see later on. In this chapter we shall study some common flowers belonging to two two-lipped, or bilabiate families.

The Mimulus, or monkey-flower, Fig. 52, belongs to one of these families. The kind of Mimulus in the picture grows on a shrub several feet high. In many parts of California there are hillsides entirely covered with these shrubs, and it is worth going far to see them when they are in full flower in late spring time. They bloom most profusely, and the flowers are large and beautifully colored; sometimes pale yellow or salmon, sometimes buff or orange, sometimes quite red. There is another very common Mimulus, which grows along streams or irrigating ditches; it has large yellow flowers with spotted, nearly closed throats, the shape of the flowers suggesting the common name, snapdragons.

Now both kinds of Mimulus have exclusive flowers. The shrubby Mimulus has corolla tubes two or three inches

Fig. 52. MONKEY-FLOWER—*Mimulus glutinosus.*

long, and so invites only humming birds and the largest moths. The other Mimulus has a shorter corolla, but by having the throat nearly closed and covered with dense hairs, it keeps out many small insects. One would expect such exclusive flowers to proffer abundant hospitality to their chosen guests, but they are much more showy than generous; they serve little honey, and bees and humming birds are quick to learn this. In the day time one rarely sees the paler flowers of the shrubby Mimulus visited, but perhaps the night moths pay them frequent visits; they could find them easily. The red flowers seem to be visited by humming birds more frequently than the other varieties; perhaps where the red flowers prevail, it is because they are near humming birds' haunts. You will occasionally see large bees forcing entrance through the closed throat of the other Mimulus.

But both kinds of Mimulus have excellent devices for making the most of a few visits. Close against the upper lip, or roof, of the flower lie the stigmas and anthers. The stigma consists of two lips, which lie open if the flower has not been recently visited; touch the lips, and they promptly close; so, if the stigma is touched by a guest's head with pollen on it, what is left there is securely shut in. The anthers are peculiar; by looking closely you will find that there are really four of them, each with two cells; they lie close together, quite below the stigma. Press against them and see how they open and leave pollen on your finger, just as it would be left on the head of a large insect or of a humming bird crowding into the flower to reach the honey. Notice that the Mimulus needs but four stamens and one pistil. You will find that all plants of high rank are economical. The members of these bilabiate families have rarely more than four stamens, and many of them succeed with only two.

The Collinsia, Fig. 53, sometimes called innocence, belongs to the same family as the Mimulus. It, too, has a bilabiate corolla, four stamens and one pistil. The lower lip, of course, consists of three united petals, but the middle petal is folded so that it resembles the keel of the lupine; in fact, it serves the same purpose; it enfolds the stamens and pistil, protecting the pollen and stigma, but when a bee alights on the lower lip, the fold spreads enough to allow pollen or stigma to rub against its body. The

Fig. 53. INNOCENCE—*Collinsia bicolor.*

Collinsia does not furnish so much pollen as the lupine, but the supply lasts a long time, because the anthers shed pollen one after another; besides, the Collinsia provides honey, saving it for the bees, and advertising it by pretty spots and lines. The stigmas are mature in the older flowers.

Another member of this family is the owl's-clover, or the painter's brush, Fig. 54. The pink or purplish and white flowers, grow together in a very social way, and the little leaves or bracts among them have their tips white or pink, thus helping to make the cluster showy. Each flower keeps its honey at the bottom of a tube too deep and narrow for bees and small insects, yet it dares to keep its stigma, which is like a fuzzy ball, quite above its anthers; that is, the flower trusts entirely to its guests for pollination. Now the anthers are tucked away in the upper lip, and the queer, puffy, white, lower lip is not large enough for a platform, still some bees know how to get the pollen. They cling to the upper lip while they drag out the pollen, and so must strike the stigma. Butterflies and moths seem to be the preferred guests; and the white lower lip shows night moths the way to the honey.

The other flower in the same picture, has several common names, painted cup, scarlet painter's brush, Indian plume, etc. It is a plant beloved by the humming birds. The tips of the bracts and the calyxes, as well as the corollas, are scarlet, the humming bird's favorite color; and the honey is beyond the reach of most other guests. The lower lip of the flower can scarcely be discerned, but a platform is of no use to humming birds. You have seen them dart from cluster to cluster, pausing the merest instant at each flower as they sip its honey. No other guest is half so swift, so it is not strange that so many flowers reserve their honey specially for humming birds. When the painted cup grows in masses, one rarely fails to see the birds paying their visits;

Fig. 54. OWL'S CLOVER AND PAINTED CUP— *Orthocarpus purpurascens* and *Castilleia parviflora*.

and when there are only scattered plants, the clusters are so showy, and last so long, that the flowers can be pretty sure of a visit sooner or later; at any rate, they make no provision for self pollination.

Another group in this family is best called by its botanical name, Pentstemon. Scarlet Pentstemons are often called honeysuckles, and violet or blue ones, bearded tongues. The Pentstemon in Fig. 55, is a violet one, and there is a picture of a climbing, scarlet Pentstemon in Fig. 71. Another scarlet one, with long, slender flowers, is sometimes called the scarlet bugler; in Southern California it begins blooming in February, and its brilliant clusters last until mid-summer; the violet one, also, is in flower nearly all the year. There are many Pentstemons that bloom only in summer time, the greater part of them being mountain plants. The name Pentstemon, or fifth stamen, is given from the fact that there are five filaments, but, as you can see, one of these has no anther and so is not a true stamen; still it has its use. Pull off a corolla and look through it as you would through a telescope. You will see that the four true stamens curve toward each other, and nearly meet, forming a dome-shaped screen just above the honey, while just above this, the fifth stamen crosses from one side of the flower to the other, so leaving a very narrow entrance to the honey.

All Pentstemons, then, are exclusive flowers. The scarlet ones have tubes too long for the bees, and invite humming birds; the blue and violet flowers admit bees, but some kinds, only the largest ones. Anthers and stigmas are always against the upper part of the corolla, and are sure to be struck by a large guest. The lower lip of the scarlet flowers is not conspicuous, in fact it can hardly be made out at all in the scarlet bugler, but humming birds need no platform; the blue Pentstemons have broad lower lips, which form good platforms. There are insects that take

BILABIATE FLOWERS

Fig. 55. THE VIOLET PENTSTEMON—*Pentstemon heterophyllus.*

revenge on the Pentstemons for their exclusiveness. They find the tubes too narrow, or else they are barred out from the honey by the interlacing stamens, but they have discovered that the corolla walls are thin, so they bite through them, making neat little holes through which they, and other short-tongued insects, can sup honey. Of course this is stealing, for these insects touch neither anthers nor stigmas, and so do not carry pollen for the flowers.

The mints, hoarhound, Salvias and sages, are bilabiate flowers, and they are nearly related to the others we have been studying; but botanists put them in another family because their ovaries separate into four parts. In the valleys of Southern California, native mints are not common, and the hoarhound, being a sturdy weed, will be studied later; but the Salvias and sages are abundant, and are well worth considering now. Most of them have their flowers arranged in compact heads scattered along the stems. The flowers have but two stamens and one style.

The Salvia of Fig. 56 is a very common one on gravelly hillsides. Its flowers have deep blue corollas, but the sharp pointed bracts and calyxes are much more conspicuous, being a rich brownish red, sometimes nearly wine-colored. The Indians call the plant the little chia, and prize its seeds very highly for food. There is another Salvia, the big chia, or the thistle-leaved Salvia. You are sure to remember this plant if you have seen or handled it, because of its beauty of color and its ability to defend itself. Every point on leaf, bract and calyx, is tipped with a sharp spine, yet the whole plant is clothed with the most exquisite mantle of loosely spun, cobwebby hairs; while, standing out against this grey-green ground, are the airy lavender flowers with their delicate slashes and frills.

In Fig. 57 are two common sages of Southern California. No. 1, the button sage, or black sage, No. 2, the white sage, so famous as a bee plant. Both sages and

Fig. 56. LITTLE CHIA—*Salvia Columbariæ.*

Salvias have exclusive flowers, the corolla tubes being sufficiently long and slender to keep the honey for desirable guests. The sages make exclusiveness more certain by having tufts of hairs in the corolla tubes. The little chia invites hive bees, and is almost sure to be pollinated by them because of its peculiar stamens; perhaps you can study them out. The big chia invites only bumble-bees to its honey, but hive bees gather its pollen; both kinds of guests seem afraid to trust themselves to the delicate flowers; they clutch at all of the stamens, styles and fringes within reach, so that the bumble-bee is sometimes sprawling over several flowers at the same time; in this way, they may often rub pollen on stigmas, but they are not sure to render this service.

The black sage has surer methods, as you can easily see for yourselves. The flowers are blue, the bees' own color; they are usually of the proper size to exact service from hive bees, and, of course, from native bees of equal or greater size. The younger flowers hold their anthers just where the bee's head will strike them when he comes for honey; the anthers shed pollen downward, and as bees leave the flowers you can see that their heads are blue with pollen. The older flowers put their stigmas in nearly the same place, and as the bees enter you can see the polleny heads strike them.

The white sage has a curiously folded lower lip that seems designed to bar out all but the strongest bees, and it is very amusing to watch the bees on the flowers. Big bumble-bees skilfully unfold this lower lip, but they seem afraid to trust their whole weight to it, and cling to the flower by thrusting their legs over the two stamens. This brings the anthers against their bodies near the base of the wings. The mature stigmas of the older flowers are likely to rub against this pollen-covered surface, so bumble-bees usually pollinate the flowers. But hive bees, also, are

Fig. 57. SAGES—*Audibertias.*
1. Black sage, *A. stachyoides.* 2. White sage, *A. polystachya.*

clever enough to unfold the lower lip and get the honey, and they never touch anther or stigma. So it appears that the native California bees deal fairly with the white sage, but the immigrant hive bees simply steal from the flowers tons and tons of honey that we in turn take from the bees.

The sages provide rich stores of honey, and the humming birds know this very well; so, although they must share with the bees, they are frequently seen visiting the flowers. And how rapidly do you think they can make calls? I once counted those paid by one humming bird to the black sage; it visited sixty-five flowers in one minute. A sage with large crimson flowers is not rare in Southern California, but it is an herb with very disagreeable foliage, and it grows only in shady places, so it is not commonly known except by the humming birds; it keeps all of its honey for them, and they seem to find it readily.

The blue-curls, which has been mentioned in other chapters,—see No. 2, Fig. 15—belongs to the mint family, and if you keep a sharp lookout you are likely to find many other flowers belonging to one of these bilabiate families. This is specially true in the summer time in the mountains, where there are other mints and sages and many kinds of Mimulus and Pentstemon. It is not strange that these flowers of high rank are more common in the late spring and summer, when there are more humming birds and large insects; for probably the success of these families is largely due to the fact that they know so well how to secure ample returns for their hospitality.

CHAPTER XII.

SOCIAL FLOWERS.

Botanists do not fully agree as to the rank of certain plant families, but the first place is generally accorded to the family Compositæ, which includes the sunflower, daisies, tidy-tips, the thistle, dandelion and the like.

Find a sunflower that has been open for several days and examine the darker central part. It is easily seen that it consists of very many slender, goblet-shaped, yellow flowers tipped with brown or purple. These closely packed flowers have not a green calyx; they could have no use for one. In the sunflower, each calyx is reduced to two or three scales. The corolla has its five petals united nearly to the tip; perhaps you will find traces of honey at the base of the tube. To study stamens and pistil, you need to split open the little flower. The five, dark anthers are united into a tube, and they shed their pollen inward. The pistil consists of an ovary, which is below the rest of the flower, and one style, which is split at the top, exposing two rough, moist surfaces, the stigmas. Perhaps you can see that the upper part of the style is covered with little bristles; under the microscope it looks like a brush for cleaning bottles or lamp chimneys. So we will remember that in every tiny flower this style, like a bottle brush, fits closely in the tube formed by the anthers, and that the inner surface of the anther tube is covered with pollen.

Fig. 58. SUNFLOWER—*Helianthus annuus.*

Now look again at the entire sunflower, which you see is not a single flower at all, but a flower cluster. Lying next to the bright yellow circumference, is a ring of tubular flowers that have their pistils extending beyond the anthers, the tips curled back, exposing the stigmas; then comes a ring of flowers with the style tips just visible, or perhaps with a tiny heap of pollen on the top of the anthers; within this ring, no pistils can be seen. The story now is not hard to read. When the flowers first open, the brush-like style is still within the tube of anthers, but it keeps on growing, and, as it pushes its way through this chimney lined with pollen dust, it sweeps out the pollen before it. So every little flower first furnishes pollen for other flowers, then spreads out its own stigmas.

Now examine the bright yellow part of the sunflower. Each yellow ray is really a flower, but an imperfect one. There is a trace of ovary and calyx, but no true stigma, and no hint of anthers. The main part is the corolla, and a very peculiar one it is! At the base it is tubular, like many other corollas, but for the rest of the way it seems to have been split open and laid out flat, so it is called a strap-shaped corolla. These outside flowers are also called ray flowers. Since they have neither stigmas nor pollen, they cannot help directly in seed-making, and seem to exist only for show.

There still remains the outside circle of green about the sunflower; this, like the outside leaves, or bracts, of any flower cluster, is called the involucre. Let us notice the sunflower's devices for seed-making, or rather for pollination, which must precede seed-making. Each individual flower keeps its own stigmas folded together as it sweeps out the pollen, so it depends upon its neighbors for pollination. Sometimes the stigma must strike the pollen of neighboring flowers as the style tips turn back, but the flowers can surely depend on insects for carrying pollen, since they furnish an abundance of it, and some honey besides. Flying guests are attracted by the great yellow ray flowers, and so are likely to begin collecting pollen from the outside tubular flowers. These flowers have stigmas exposed, as you remember, and the guests bring them pollen from other sunflowers they have been visiting; this is supposed to be better for them than pollen from nearer neighbors. As the insect goes on collecting pollen from the inner flowers, fresh supplies cling to him to be carried to the next head.

You have probably collected many flowers that are, in structure, similar to the sunflower. By April, or even March, members of this family are among the most common kinds of wild flowers. The early kinds are usually yellow; they are all commonly called daisies. There is one that in April quite carpets many acres of untilled ground in Southern California, making patches of yellow that can be seen miles away; hence the plants are sometimes called golden fields. These flowers furnish considerable honey as well as pollen, and have a marked fragrance; as one would expect, they are thronged with insect guests, especially flies. Ranchmen say that these flowers, because of the flies they attract, are the cause of much suffering to horses and cattle. Among the prettiest of these daisy-like flowers, are tidy-tips, which take their name from the white tips of the ray flowers. Notice especially their calyxes as the corollas wither and the fruits mature; you can readily think out their use.

Not all members of the family Compositæ have the two kinds of flowers. You will find many of them quite without strap-shaped flowers, and others consisting entirely of them. Some of the latter are the dandelion, chicory, sow-thistle and the pretty summer weed in Fig. 12, Chapter III. Most of the flower-heads of this group remain open but a few hours each day, and the plants have milky, sticky juice, so it is rather troublesome to study them in the school room, but their fruits and the pretty calyxes that cling to them, can be easily watched.

The thistle is one of the Compositæ that has no strap-shaped flowers, and indeed it does not need flowers that exist for show alone, because each one of its tubular florets is, in itself a thing of beauty, with its fluffy calyx and slashed, lavender corolla, tipped with crimson. Like many other Compositæ, the thistle provides entertainment for two different classes of guests; there is pollen for the shorter-tongued insects, and honey served in slender cups for guests of higher rank. Most of our California Compositæ are white or yellow, the most conspicuous colors, and hence the colors most useful in attracting the lower insects; but the thistle seems to aim to please all classes by offering plenty of white pollen to the common herd, and keeping honey in blue and crimson flowers for bees, humming birds and butterflies. At any rate, these more aristocratic guests do frequent the flowers.

As the dry season advances, you will find that a large number of the plants that are able to survive, belong to this family, Compositæ. You will remember that many of our autumn plants belong here. We could trace their success partly to their ability to defend and protect themselves by means of hairs, spines, thick coats, resinous matter, and so forth; they have also many clever devices for scattering seed, but the fact that, in this family, the flowers live close together in flower cities, or communities, has probably very much to do with their success. As a matter of fact, it is by far the most successful of all plant families. Botanists tell us that about ten thousand out of the one hundred thousand kinds of known flowering plants, belong to this family.

Another large family is the Umbelliferæ. As the name implies, the flower cluster is like a little umbrella, an

Fig. 59. THISTLE—*Cnicus occidentalis.*

umbel, the botanists call it. The little flowers that make up the cluster are not of so high a rank as the florets in the Compositæ. They do not unite their petals into a cup for honey; instead of this, the honey forms a thin layer over the center of the flower, and so can be reached by the humblest guests. The color, as one would expect, is almost always white or yellow. The flowers in any cluster usually all look alike, but sometimes the outside ones are larger than the others, and rather frequently some flowers have only anthers mature, while others have stigmas ready for pollen.

Some of the earliest plants to appear after the rains, belong to this family. One of these, with its underground part, in shown in Fig. 22, Chap. V. Its rather homely little flowers are out by January, and several kinds of similar Umbelliferæ are abundant all through the winter and early spring. By May their fruits are mature; some of these are like the fruit in Fig. 60; you can see how the seeds hang by delicate threads until a wind scatters them. There is another group, with very prickly, or bur-like, little fruits, which you have probably helped to distribute by means of your clothing.

During the later spring and early summer months, the wild celery, hemlock and several other vigorous Umbelliferæ, flourish in damp places. In the towns of Southern California, one of the most common wayside summer weeds is the fennel, another member of this family; it has very delicate, feathery foliage and yellow flowers. The plant is strongly scented, as are many other Umbelliferæ, seeds of which you probably know, such as caraway, coriander, anise and dill. The pungent taste must protect these from the attacks of animals. Many Umbelliferæ are actually poisonous both to man and animals.

There are many other plant families, of various ranks, that have adopted the plan of massing small flowers into

Fig. 60. AN UMBELLIFERA—*Peucedanum utriculatum.*

communities, and the device always seems to be successful. So let us consider for a moment the advantages of flowers in cities, over solitary flowers or flowers in small clusters.

You have read perhaps, or have heard other people tell, of times when each man had to make his own house, or fort, or castle, and defend his family as well as he could against enemies. He was obliged to make his own tools and weapons, raise his own food, and make his own clothing from wool, flax and leather that he, himself, provided. But as a country or a race becomes older and more civilized, as we say, the defense of the people is left to one class of men, the police or soldiers; another group of men raise food; others manufacture tools, furniture, etc.; so each man has a chance to do one thing well. In civilized countries, too, there are some people who seem to take no part in the toil of the community, the leisure classes, we call them.

In our sunflower or thistle community, the involucre is clearly the defensive part; besides, it may aid the rest of the plant in food-making. Since the involucre takes the place of individual calyxes, the calyxes may be developed for another use; so we see them becoming floaters or hooks of some kind, thus aiding in seed distribution. Some of the little flowers furnish pollen, while others have stigmas ready for its germination, and still others, like the ray flowers of the sunflower, serve simply to attract guests. It is possible that our leisure classes, corresponding to ray flowers, may also be of some use to our community. Another advantage of flower-cities is, that any good that comes, like the insect visits, can be easily shared by all; just as men, congregated in cities, can share, with the many others, good schools, good music and other city advantages.

CHAPTER XIII.

PLANT FAMILIES. PART I.—ENDOGENS, OR MONOCOTYLEDONS.

By a plant family we do not mean a group of plants that have the same parent plant, but those that resemble each other in other certain ways; just as among animals the lion and tiger belong to the cat family, or the wolf and coyote to the dog family. Arranging plants in families is called classification. It is true that the best botanists do not fully agree in their way of classifying plants, still there are many relationships that are easily seen, and it is a pleasure to note them; beside, one needs to know something about plant families in order to use books about plants.

The most important thing a plant does, is to produce other plants; so its way of doing this usually counts most in classification. You remember that sea-mosses, mould, toadstools and even ferns, grow from spores, and that spores are simple plant cells, subject to very many dangers. But nearly all the plants people commonly see and think of as plants, produce seeds; and a seed, as we have seen, instead of being a mere cell, is a little plant, which has been formed and provided with some food before leaving the parent plant. So the whole plant world is divided into two great groups, seedless plants and seed-bearing plants; these groups are divided into smaller groups, which are again divided, and so on, long Greek or Latin names being given to all the groups.

ENDOGENS OR MONOCOTYLEDONS

Seed-bearing plants are first divided into those that do not enclose their seeds in an ovary and those that do. Pines and other cone-bearers belong to the lower, or naked-seeded group, as it is called. But it is only when the cones are very small that the seeds, or rather, the ovules, are uncovered; they then lie where the pollen can reach them directly, there being no ovary, style, or stigma; but the scales of the cone soon close about the developing seeds, and in the end they are better protected than most other seeds. So, unless you are a botanist and study the reproduction of plants with the microscope, it will hardly seem just to you to rank the sturdy group of cone-bearers lower than many insignificant little weeds.

Perhaps you have seen the pines rising above the winter snows on our mountains, and have read of the great stretches of pine forests in cold north countries like Norway or Siberia, or the colder parts of our own country. In fact, cone-bearers can generally brave cold better than drought, and in many parts of California they do not grow without irrigation. The Monterey cypress is a cone-bearer much used all over California for hedges and windbreaks; the trimmed trees are stiff and uninteresting, but near Monterey the cypresses grow naturally on the wild, rocky coast, sometimes where the salt water splashes them, and artists love to picture these sturdy old trees with their knotted, twisted limbs, draped with long grey lichens.

The California "big trees," too, are cone-bearers; they are famous the world over, and you probably know some wonderful stories about them. They are a kind of redwood, and are found only in the Sierras, but there is another kind of redwood that grows in the magnificent forests along the northern coast of California; the story of how the lumbermen cut, haul, and saw these trees is very interesting. The northern Pacific coast has, too, some of the finest pine trees in the world. Can you think of some

CALIFORNIA PLANTS IN THEIR HOMES

special uses for the timber from very tall pine trees? Think whether the family of cone-bearers furnishes a considerable part of wood used in building and manufacturing. You will decide that, whatever rank the botanists may assign to this family, it is certainly a very interesting and useful one.

The other division of seed-bearing plants, the division that produces seeds in ovaries, is divided into two great classes. To illustrate the first class, collect all the following that you can, remembering to get roots, leaves, flowers and seeds when possible:—lilies and plants resembling lilies, such as Yucca, gladiolus, crocus, flag; different kinds of grasses, including wild oats, wheat, barley and Indian corn; callas, cannas, cat-tails, rushes and tules. Perhaps some one can furnish the flowers and young fruits of the banana, palm and century plants; many of you can at least watch these last three plants out of doors.

The seeds of plants of this group are usually difficult to examine, but they all, like the corn and onion seeds, have only one seed-leaf or cotyledon, so the plants are called monocotyledons. Cut thin slices across the stems of these plants, and notice how the woody strands are arranged. They are more compact near the outside, but there are some scattered strands all through the softer central tissue, or pith, as it is called. Stems that have the woody strands arranged in this way are called endogenous, and so plants of this group are endogens. This is not the most economical way to arrange the wood; the very best way is for stems to have all of these woody strands form a cylinder about the pith. Most grasses omit the pith entirely except at the joints, and so a wheat straw can support many times its own weight, and can bend with the winds without breaking.

Now notice the leaves of this group of plants. Are they usually simple or compound? Are their edges notched and cut, or entire? Are the woody strands, or veins, parallel or netted? How are they arranged in the buds that are unfolding into new shoots? Count the number of parts in the flower. What seems to be the most common number? Examine the underground parts. How many of them have storehouses? Perhaps now you can see some reason for putting these plants in the same group.

One division of the group of endogens, is made up of the lily family and nearly-related families. Recall what you found out about the cluster-lily, Fig. 33, and the Mariposas, Fig. 48. California has many other wild lilies, some of them so beautiful that they are grown in gardens and much prized in foreign countries. There are the brown lilies that come early, and there are other Brodiæas besides the cluster lilies; the true wild onion has very pretty flowers, and is common in many places; then in the mountains there are several kinds of large lilies, some of which, like the Humboldt, or tiger lily, the leopard lily and Parry's lily, come in the summer months. In May children in Southern California may be able to get the Yucca; also some smaller, bright yellow lilies that grow in large umbels, and are sometimes called golden star lilies.

Collect all the kinds of wild lilies you can find and look also in your gardens for plants with flowers that look like lilies. From these and from pictures, you will find that the true lilies have always three petals, three sepals, twice three stamens, and a three-celled ovary that is called superior because the other parts start below it and are not united with it. Now any flower having this sort of an ovary and the other parts in these numbers, belongs to the lily family; remembering this fact will help you in finding from books the names of flowers new to you. In gardens, you are likely to find some very pretty flowers with parts in threes, but with inferior ovaries, so that the seeds seem to be in the stem; botanists put these flowers into two other families; the Amaryllis family, which includes the Chinese lily, Freesia, century plant, snow drop, narcissus, jonquil and the like, and the Iris family, which includes, besides the Iris, or flag, our blue-eyed grass, Fig. 61, the gladiolus, and crocus. The Iris family has three stamens, the Amaryllis has six. You are likely to find some very stately true lilies among the garden flowers; some of the most elegant of these foreigners come from Japan. The pretty South American climber that we call smilax, belongs to the lily family; so do the hyacinth, red hot poker, tulip, and the more humble but useful onion and asparagus. Men find uses for some other members of this group. In Mexico, where the century plant grows abundantly, several kinds of intoxicating drinks are made from its sap, the strong wood fibres are made into ropes, and the underground stems furnish material for soap. The stems of one California Yucca, and the bulbs of our soap-root are also used as soap. Some members of the group furnish medicines. The rushes, which are nearly related to the lilies, are used for weaving chairs, baskets, etc.; the pith of their stems was formerly used for wicks; perhaps you have heard old people tell of rush lights.

On the whole, from our standpoint, this group seems to be more ornamental than useful; but we must remember that for the flowers, beauty is use, and that these flowers with their beautiful forms, colors and fragrance will furnish us many wonderful stories, if we have the wit to read them. The most wonderful story of all, perhaps, is that of the Yucca. This plant is one that knows how to meet hard times; it can gather and store moisture well enough to flourish in the desert; it makes and stores so much food that when the right season comes, there shoots up, in a few days, a stem sometimes twelve feet high, bearing one of the most beautiful flower clusters in the world. The leaves can defend themselves so well that some kinds of Yucca are called Spanish bayonet; and yet the plant depends for the

ENDOGENS OR MONOCOTYLEDONS

Fig. 61. BLUE-EYED GRASS—*Sisyrinchium bellum.*

most important thing of all, its seed production, on a tiny moth no longer than your finger nail. The Yucca keeps its anthers quite away from its big stigma, and does not even provide honey for the moth; but the white, fragrant flowers are easily found at night. The moth herself does not need food, but she knows that when her eggs hatch, her little larvae will be very hungry indeed, and the only food they can live on is the Yucca seed. But she seems to know something much more wonderful than this, and that is, that the Yucca will not produce seed unless the pollen reaches the stigma. At any rate this is what she does: Before she lays her eggs, she gathers a mass of pollen bigger than her own head; she next lays her eggs in the ovary of a flower, and then immediately crams the pollen down the concave stigma of this flower. So the big Yucca can trust to the mother instinct of this tiny creature, for while her babies eat part of the seeds, there are quite enough left to keep up the supply of Yuccas.

Another group of monocotyledons is made up of families so different in appearance that you would never guess their relationship. It contains, for instance, the largest of all monocotyledons, the palms, and also the smallest of flowering plants, the duckweed, a tiny water plant hardly as large as a common pin. There are several other water plants that belong to this group, flowering plants that either float on the water or grow along the margin of ponds. Perhaps the one you would know best is the cat-tail, or bulrush. But the best known of this group is the calla, which you have already studied. All these plants are related because, like the calla, they have small simple flowers, usually many of them crowded together. The clusters differ almost as much as the plants. There may be a smooth spike surrounded by a showy involucre like the calla, or a brown, velvety one like the cat-tail, or a much branched but compact cluster, as in the palms, or one so

Fig. 62.—*Yucca Whipplei.*

inconspicuous that you can hardly find it, as in the pond weeds.

The palm family is by far the most useful one in the group. In our climate we cultivate palms for ornament, but in tropical countries they furnish food, drink, clothing, houses, boats, furniture, utensils, ornaments,—in short everything the people of these countries require. You can find out many interesting things about the most useful of these trees, the cocoa palm, date palm, sago palm, etc., in text books on Physical Geography or in Encyclopedias.

By far the most useful group of endogens is the one that consists of the grass family and the nearly-related sedges. When you know that the grass family includes wheat, rice, oats, barley, corn, sugar-cane and bamboo, you will admit that, of all plant families in the world, this must be the one most useful to man. In tropical countries some grasses become almost trees, the bamboo, for instance; the cooler countries furnish many of the most valuable of this group, and there are some grasses, such as the bunch grasses of our western plains, that can adapt themselves to drought as well as to cold.

But in spite of all differences in size and habit, there is a strong family resemblance among the grasses. What seems to be always true of the stems of the leaves? The flowers you have, perhaps, not recognized as flowers, because they are green and inconspicuous, but if you think for a moment, you will know that the heads of wheat, the ears and tassels of corn, and the plume-like tops of the oats, are all flower clusters. You can find the stamens and pistil of the flower, and perhaps you will have patience to make out the wrappings. Of course the absence of bright color means that this family does not ask for insect help, but trusts entirely to the wind for pollination See how the ripe anthers hang, like banners, from slender stems so that every breath of wind shakes out pollen; then notice the

Fig. 63. WILD OATS—*Avena fatua.*

feathery stigmas that serve so well for catching pollen. The stigmas of the corn are the lines of little teeth that extend the entire length of the styles, or silks. And so the grasses get well pollinated, and produce enough seed to keep up the supply of plants, and feed countless men and animals besides. In our climate, some introduced grasses, like wheat and corn, need artificial help, that is, they must be cultivated; but there are many grasses, native and cultivated, that can take care of themselves very well indeed. Find some of the Bermuda grass that infests the lawns, and see why it spreads so fast. What helps to scatter the foxtail grass and the wild oats? Notice the bristles of these grasses; run them up and down between your fingers; can you see the use of the little barbs on them? Put a long bristle of the wild oats in water, then take it out and let it dry in the sun; wet it again. What do you think can be the use of the movements it makes? Find other grasses with bristles of like habits.

The sedges grow in moist places, and resemble the grasses. What is commonly called the tule, in California, is really a sedge, so is the Papyrus of our gardens. Tules, as you probably know, are used to build huts or to thatch adobe houses, and the Papyrus was once used for paper; but generally the sedges, like their neighbors the cat-tails and rushes, are of more use to birds and water-fowl than to man. They often defend themselves against cattle by sharp teeth along the edges of the leaves.

A higher group of endogens contains the pine-apple, ginger plant, canna, and banana. Perhaps you have seen the banana bearing fruit in California; in tropical countries great quantities of fruit are produced; in fact, bananas form the principal food of millions of people. The bright flowers of the canna are specially interesting; try to make out the parts for yourselves; call the three, small, outside parts the sepals, and be sure to find the stigma and pollen.

You will discover that the stamens and pistil are like petals and are scarlet also, and that the flower can get along with one one-celled anther. Can you guess from the color what carries the pollen? You probably would not have to watch long to see the flowers visited by the most desirable of all guests.

The orchid group is one of the highest, and perhaps quite the most interesting, of endogens; but the orchid family, although a very large one, does not find the California climate congenial, and we have very few varieties. Perhaps you have seen orchids in greenhouses or in florists' windows, or you may have heard or read about them; of how they sometimes grow among the tree tops of dense forests, or of the curious forms of the flowers. Darwin wrote an entire book about orchids, and it contains many wonderful stories. There are orchids that trap their insect guests and make them carry pollen in order to escape; there is one kind that actually hurls masses of pollen at bumble-bees' backs, where they stick until left on the stigma of another flower; another kind fastens bags of pollen on its guests' eyes. This last kind is not rare in California, though its flowers are not so large as those of the English species that Darwin writes about. It is an interesting plant, and will amply repay any one who has the patience and time to study it.

CHAPTER XIV.

PLANT FAMILIES. PART II.—EXOGENS, OR DICOTYLEDONS.

How many seedlings can you recall with two seed-leaves or cotyledons? Aid your memories by turning to Figs. 5, 6 and 8 in Chapter II. These, and like plants, are called dicotyledons. Find some of them older grown, and cut across their stems to see how their woody strands are placed. Plants with this arrangement of wood in the stem, which is the best possible, are called exogens. How are the strands, or veins, arranged in the leaves? Compare the leaf of the castor bean with that of the corn, for instance. Which do you think the stronger? Which group, endogens or exogens, has the greater variety in form of leaves? Do you remember why some plants are better off because of their slashed or divided leaves? Because exogens, or dicotyledons, as a class, are better able to meet dangers than endogens or monocotyledons, they are considered of higher rank, but it does not follow that they are more useful to man.

Many botanists put in the lowest group of dicotyledons those that have small, incomplete, and, usually, crowded flowers. The willow, walnut, sycamore, alder, oak and fig belong here; so do some other useful trees, which, perhaps, you do not know; other nut-bearing trees, such as the chestnut, pecan and beech, and shade and timber trees, like the elm and birch. There is the mulberry, which feeds

EXOGENS OR DICOTYLEDONS

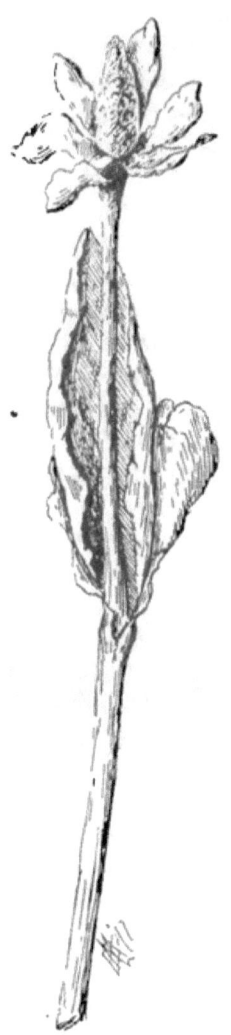

Fig. 64. YERBA MANSA—*Anemopsis Californica.*

silkworms, and its cousin, the bread-fruit tree of the Pacific Islands, and there are some interesting and useful relatives of the fig, one of which you may find out by yourselves because of its fruits; it is the India-rubber tree of our parks; you can guess what part of the tree becomes the rubber. Another member of the fig family is called the cow-tree, because its juice is like milk and is good to drink.

Some members of this group are not trees. There is the yerba mansa, Fig. 64, which is also called alkali weed. What seems to be a flower, is really a cluster of small, closely-packed flowers, as you can see for yourselves. Many people think the yerba mansa useful as a medicine; it has a relative in India whose ground berries we call black pepper. One family in this group is troublesome rather than useful; that is the nettle family. You probably think our own nettles bad enough, but in Australia, nettles become trees, and their sting is very severe.

Next to this group comes one we might call the weed group, because it contains so many of our common weeds. We shall study the weeds in another chapter, but some of the weediest of families have members that have been cultivated to usefulness; the beet and rhubarb, for instance, are nearly related to the dock and knot-weed. Other weedy families also have members noted for beauty; chickweed and purslane are troublesome weeds the world over, but the chickweed belongs to the same family as the carnation pink and our beautiful, wild Indian-pink, and purslane is a member of the Portulaca family. The wild four-o'clock and the sand verbena are classed in this group, but it is not easy to explain why. There is one useful family in this group, the buckwheat. The cultivated buckwheat not only furnishes flour, but is also a valuable honey plant, and some kinds of our wild buckwheat, or Eriogonums, are also useful to bee-keepers. The Eriogonums can defy drought, as you remember, and so it happens that in the summer months they convert hot sandy or rocky wastes into fragrant bee pastures.

The higher groups of dicotyledons consist mainly of plants with more conspicuous flowers, that is, they have learned to get help from insects, and the highest groups of all have united their petals so as to get the most possible good from their guests. The group next above the weedy group has flowers with little or no union of parts. Many of those with their parts all quite separate, belong to the buttercup family, and you will think the family resemblance very slight when you know that it includes, besides the buttercup, the clematis, meadow-rue, larkspur, columbine and peony. If you can collect the fruits of this family, you will understand more about their kinship. The magnolia and laurel families, too, have flowers with their parts not united. Perhaps you know the laurel, or bay tree, of our cañons; you are likely to miss its small greenish flowers because they come so early in the winter, but you will remember its beautiful, dark, glossy leaves, especially their spicy odor and taste, and you will not be surprised to learn that cinnamon, nutmeg, mace, cassia buds, camphor and sassafras all come from plants nearly related to the bay-laurel.

Some of the wild flowers noticed in Chapter IX. belong in this group; the violet, poppy and mustard and all their relatives. How many relatives of the mustard have you found? Shepherd's purse and pepper grass, if you have kept a sharp lookout on wayside weeds, and water cress along streams, sweet alyssum in gardens, radish, cabbage and turnip about old vegetable gardens, besides a goodly number of wild flowers, including the wallflower and others not showy enough to have common names. This large and vigorous family, then, contains some members that seem to us quite useless; others, the weeds, that are worse than useless; but some that are ornamental, and still others decidedly useful. The sharp, biting taste, which must protect the plants from grazing animals, is agreeable to most people; that is, we use mustard, water cress, radishes and horse-radish as relishes. The original wild cabbage has been cultivated into not only various kinds of cabbage, but also kail, cauliflower, Brussels sprouts, and other garden vegetables. All kinds of turnips are supposed to be cultivated forms of a common weed known as the smooth-leaved mustard.

One of the highest families in this group is the malva. The flowers of the common weed, malva, are easily missed because they are so small and close to the stem, but the fruits are well known; the children call them "cheeses", and like to eat them because of their peculiar gummy or mucilaginous juice. But small flowers are by no means the rule in this family; it includes the hollyhock, hibiscus, flowering-maple, and many showy wild flowers, some of them growing on shrubs or even trees. Find any of these you can, and look for family traits; what is always true of the stamens? How many have mucilaginous juice? Have you never used the juice of the hibiscus flowers to polish your shoes? Perhaps you are fond of the vegetable okra, or gumbo. Can you guess why it belongs to the malva family? The most useful member of this family is the cotton plant. It is quite possible to grow cotton in California, and you may be able to find out many interesting things about it.

This group of dicotyledons contains also many plants that you are not likely to meet, though they are noted for beauty or use, or have some peculiar interest; the water-lily, for instance, the tea plant, the chocolate tree and the curious pitcher plants and sundews, which thrive on animal food. Interesting stories have been written about them all.

The next group of dicotyledons contains families whose kinship is not easily seen. The flowers, of course, count most in determining relationship; their parts are nearly always in fives, the stamens never more than twice five; the flowers may be showy or fragrant, appealing to insects, like the geranium, orange and wild lilac, or they may be small and green, depending on the wind, like the grape and poison oak. Some of the families consist only of herbs, others contain shrubs, vines and trees. You remember some of the plants that are related to the geranium; flax is also nearly related; perhaps you have seen its pretty, blue flowers, and know what use is made of the strong woody fibers of the stems. You can think of two or three relatives of the orange. The poison oak belongs to the Rhus family, which includes many other California shrubs, most of them with evergreen leaves; some kinds are known as sumach, and their leaves are often very brilliantly colored when about to perish. The cultivated Virginia creeper, a near relative of the grape, has also gorgeous autumn coloring. The shrubs called buckthorn are nearly related to the California lilacs, and so belong to this group; also the pepper tree of our streets and the maple, buckeye and horse-chestnut.

The next group contains one of the largest and strongest of plant families, the Euphorbia family, but most of its members are tropical. We have introduced some of the more hardy of these foreigners, the Poinsettia, for instance, the shrub that makes such a gorgeous display of scarlet about Christmas time. Possibly you have noticed that the Poinsettia has very milky juice, and that the scarlet part is really a whorl of leaves surrounding a cluster of small, simple flowers. There are a few California Euphorbias. One of them, called rattlesnake weed, is common in the south; it grows in round mats close to the hard sun-baked earth and, like the Poinsettia, it has milky juice and flower clusters surrounded by bracts, but the bracts, which are brown, with white margins, are very small, in fact the whole flower cluster is not much larger than a pin head. The turkey-weed, too, belongs to the Euphorbia family, and so does the castor-oil plant, but you would hardly see why. A family trait of the Euphorbias is milky or poisonous juice. One Euphorbia is sometimes cultivated in California because gophers are supposed to be poisoned by eating it. A large part of the India-rubber used in the world is made from the juice of South American trees of this family. Another South American member of the family is the Manioc, which, though it has a poisonous juice, furnishes the tapioca of commerce, and besides, is the chief article of food of the natives.

The highest group of dicotyledons with petals not united, is called Calycifloræ, because the calyx, or the calyx and receptacle, usually form a cup about the ovary. When the ovary is grown fast to the cup or tube, it is called inferior, because it is below the rest of the flower. This group contains several families that we have already studied, and some others common in California. The Umbelliferæ family, which we studied under social flowers, is one; another is the saxifrage family, which includes currants and gooseberries. Do you remember how early in the winter these saxifrages sent out new leaves and flowers? To be able to flourish with little warmth seems to be a family trait, and most of the California saxifrages are found in shaded cañons or on high mountains.

The cactus family, another member of this group, knows how to defy dry weather. Think out its devices for doing this, and find out for yourselves, if possible, a special trick the flowers have for getting their pollen carried. The ice plant, and the Sedum, or live-for-ever, both members of this group, also know how to store water.

Perhaps you remember a spring flower with a very slender inferior ovary; Fig. 42, Chapter IX, will recall it. Have you discovered any of its relatives? Examine the white or the yellow evening primrose, or the cultivated fuchsia. During the late spring and summer months, there are some very showy wild flowers that belong to this evening primrose, or fuchsia, family. Find them if you can; Fig. 69, Chapter XV, will help you. Try to find out for yourselves, before you read about them, some of their devices for pollination.

There are many interesting foreign families that belong to this group, and we can cultivate some of the plants in our climate. Some of these foreigners are:—the passion-flower vine and begonias from tropical America, the crape-myrtle and pomegranate from India, the grevillea and Eucalyptus from Australia, guavas from Brazil and West Indies; also the trees that produce cloves and allspice, growing only in the tropics, and the Brazil-nut tree in South America. You can find the flowers or fruits of many of these; notice the ovaries especially. Cloves are dried flower buds; allspice consists of dried berries. The Brazil-nuts are, of course, seeds; do you know how they are packed away in fruits that look like cannon balls? You can sometimes see them at fruit or grocery stores.

Two other families of this group are the rose and the pea families, whose members are widely scattered and very well known because of their use or beauty. The rose family includes, besides all kinds of roses, many of our cultivated fruits:—apples, pears and quinces, which belong to the same branch; peaches, plums, cherries and all the other stone fruits; and strawberries, blackberries, raspberries and the like. Collect as many of the flowers and fruits of these as you can, taking either wild or single roses,—double roses, however beautiful, are not natural. Now try to make out family traits. It is easy to make a general statement about sepals, petals and stamens, but there are many kinds of pistils. One branch of the family has inferior ovaries; the fruits will show the withered remnants of the rest of the flower above the ovaries and the surrounding calyx and receptacle. How many ovaries are there in each fruit? What part of the flower do we eat, and what do we throw away? Another branch has flowers with but one ovary, which is quite free from calyx, and the ovary walls become thick and juicy as the fruits ripen. But many of the rose family have flowers with a great number of pistils. We think of strawberry pistils as seeds because they are so tiny, but if you look closely, you will see the styles as well as the little, hard ovaries. The part of the strawberry flower that we like to eat is, evidently, the receptacle. Are blackberry and raspberry pistils and receptacles like those of the strawberry? Has the rose one or many pistils? Find some rose fruits, or hips, as they are called. What part of the flower is the soft, scarlet tissue that the birds like to eat? Have you thought out why all these members of the rose family have some part of their fruits showy and well flavored when the seeds are ripe? Why should they be green in color and disagreeable to taste before this? Of our native members of this family, some have edible fruits. We have wild cherries, wild plums, wild almonds and various kinds of so-called berries. There are some of this family that choose to have the wind carry their seeds. Did you ever see the fluffy fruits of the California mahogany tree? They are so abundant that mountain slopes covered with these trees look almost gray in the summer and autumn. The greasewood, too, whose plume-like flower clusters whiten our foot-hills in early summer, has dry fruits.

The highest of all this group is the pea family, Leguminosæ. We have already noticed one branch of this family, the one having flowers with mechanical genius. Naturally flowers of this sort, in adapting themselves so cleverly to insects, take on bright colors or are fragrant, so some of our most attractive garden flowers belong to this group. How many have you found? Another branch of Leguminosæ contains the California Judas tree, and a third branch includes the Acacias. The Leguminosæ is one of the useful families; sometimes we use the food the plant has stored in the seeds; you can think of examples among cultivated plants. On the California deserts there is a shrubby Leguminosæ, the mesquite, whose seeds the Indians eat. Sometimes it is the foliage that is used for food; the clovers and alfalfa are examples. Some Leguminosæ are shrubs or trees valuable for timber. Gum arabic is produced by some kinds of Acacia.

The flowers of Leguminosæ have usually some of their petals partly united. The flowers of all the higher groups of dicotyledons have all their petals united into a tube, at the base at least. One of the lowest of these groups is called the heath group, because the heath, or heather, of Europe belongs here. It includes some very large and interesting families, but the California members live mostly in the mountains or forests. Some of these are the manzanita, the madrone, the crimson snow-plant, the azalea and the rhododendron; cranberries, huckleberries and the like belong in this group, also the pretty trailing arbutus of the Eastern States. To another group belong the true primrose that we read about in English books, the Chinese primrose and the cyclamen that we cultivate in pots, also our own shooting star and the sturdy little pimpernel. Another group is mainly foreign; the ebony and persimmon belong to it.

The next group we have already studied. It includes the blue-eyes, Gilia, forget-me-not, nightshade and morning-glory families. Watch for the late-flowering members of these families. There are several late Phacelias, and very pretty Gilias are found in the foot-hills and mountains during the summer months. The dodder is in flower all summer, and that big, coarse cousin of the nightshade, the "Jimson" weed, unfolds its huge white flowers in the summer and autumn.

The next group of plants, too, we have studied, calling them "plants of high rank." These families, also, send out some of their most brilliant flowers during the summer months. You will find them in the mountains, or on shaded slopes and along streams in lower altitudes.

In the next group there are some common cultivated plants, the olive, jasmine and lilac, but not many native Californians. There are the gentians, which are found only in mountain meadows, and milkweeds, which are rather abundant in parts of Southern California. The milkweed flowers are not very showy, but they have plentiful stores of honey, and the story of their pollination is one of the most wonderful known. If you have sharp eyes and can watch the insects visiting the flowers, you can find out part of the story for yourselves.

There is another group of about the same rank that has few California members. It includes the cucumber family, the melon, pumpkin and the like, belonging here, also our wild cucumber vine and the chilicothe.

The highest group of all is called Aggregatæ, which means that the flowers are crowded together; and the highest family in this group, and hence the highest of all plant families, is the social family Compositæ, which we have already discussed. Does it seem strange to you that this family with its ten thousand species, has not one that is of any marked use to man? For this reason it receives little assistance from him. In fact some species, like the thistle and cocklebur, flourish in spite of constant war against them. Of course this means that this is the largest and strongest of families, simply because its members have learned so well to take care of themselves.

CHAPTER XV.

SOME SUMMER FLOWERS.

There are some plants that, all through the winter rains and spring sunshine, refuse to send out their flowers; but after the rains are over, and when the days are so long that the sun drinks up much of the moisture left in the soil, they suddenly burst into bloom. Should you not think they would pay dearly for their tardiness? But they seem to have reasoned in somewhat this way:—After the rains are over the pollen will be in less danger, and in the long sunny days and warm nights, there will be more insects or humming birds to carry it; and, best of all, when the crowds of spring flowers are gone, the late flowers will have all these guests to themselves; so it pays to store up food and moisture to be used later on.

And what a long time some of these plants have been working and hoarding! Do you remember how early last autumn the soap-root sent up its pretty crinkled leaves? All through the rainy season the busy leaves make food and send it down to the bulb storehouse, until finally the air and soil become so dry that they can work no longer. But after the leaves have quite disappeared, there shoots up a slender branching flower stalk, two or three or sometimes five or six feet high. This stalk and the unopened buds are so slender and gray that during the day there seems to be but the ghost of a plant, but watch on some June or July afternoon, about half-past four o'clock, and you will see several slender white lilies burst out on every branch. As the darkness comes on, they are like white stars; they are fragrant and full of honey, and are sure to attract their chosen guests, the night moths.

There are larkspurs, too, that provide underground stores for late flowers and fruits. In the foot-hills of Southern California there are areas fairly aflame with scarlet larkspur. The gorgeous flower clusters are nearly as tall as a man, and the oddly shaped flowers have, as you might guess, a story to tell. If you are unable to get the scarlet flowers, you can read nearly the same story from other larkspurs, wild or cultivated. The one in the picture is a native blue kind that blooms in May or June. The five outer parts of the flower are, of course, sepals; within are the four, small petals. Both sepals and petals contribute color, but they do more than this; in the blue larkspurs, the three lower sepals form a platform, and in all species the two lower petals form a roof over the stamens. The two upper petals are usually of a color different from the others, and serve as honey guides.

Now see how well the larkspur guards its honey. The two upper sepals are prolonged to form a deep, slender cup, called a spur, whose tissue is sometimes very tough and wrinkled. Really this spur, made of sepals, is only a sheath, or covering; the real cup that holds the honey is within it, and is formed from the two upper petals prolonged and curiously fitted together. And the larkspur does well to provide these thick and double walls; for, as we saw in the case of the Pentstemons, there are thieving insects that do not hesitate to bite through the flowers to steal honey, if they can get it in no easier way. Now the larkspurs invite only guests with long tongues. The blue larkspur offers honey to large bees, and it is supposed that the flowers have attained their blue color because that is the color the bees choose. The scarlet larkspur keeps its

Fig. 65. LARKSPUR—*Delphinium Parryi*

CALIFORNIA PLANTS IN THEIR HOMES

honey quite too deep for bees, but there is no doubt that its gorgeous color attracts the most desirable of all guests, the humming birds. Now look at the stamens. Each flower has many of them, but only a few are shedding pollen at any given time; note just where these stand and imagine what happens when the bee, resting on the platform, or the bird, rising from below, reaches over to get the honey. For mature stigmas, look in the oldest flowers. Where do they stand? It is easy to see what results to these older flowers every time they are visited by guests from the younger ones.

There are summer flowers, less showy than the larkspurs, that are able to trust entirely to their guests for pollination. Probably the most interesting of all these is the milkweed. The milkweed in the picture is common in Southern California. Its very milky juice, also supposed to be poisonous, and its extreme woolliness, protect it well from grazing animals and from drought. It flourishes in many waste places, sometimes to the annoyance of beekeepers, as we shall see. The flowers of this milkweed are massed in rather large clusters, but they are not highly colored. Unless you look closely, you will not find the sepals at all, and the whitish petals turn back close to the stem, as if determined to be of no use to guests. But there are five hood-like cups for honey, which are more or less deeply rose tinted; they are conspicuous, and have a goodly amount of honey.

The stamens and pistils are very queer indeed, but it pays to make them out. Tear away sepals, petals and honey cups. What is left looks like No. 2 in the picture, but is much smaller. Now explore this with a pin, and you will soon find that there are five narrow openings, or slits, with stiff, projecting edges. When a bee or a fly or a butterfly comes for milkweed honey he clings to this central part of the flower, since he cannot alight on the petals.

Do you not see how easily one of the insect's legs may be caught in a slit? As he tries to rise and free himself, the leg will be brought against the roof of the little chamber into which it has slipped, and strangely enough this roof is the stigma. One might easily fail to make out just what is stigma in this case, but the ovaries and styles of the two little bottle-shaped pistils within the chamber are easily seen; and, as a matter of fact, the upper part of the chamber is moist and ready to receive pollen. If, then, the insect's leg has pollen on it, the pollen will be landed on the stigma.

But where does the flower keep its pollen? Notice a little black dot above each slit. As the insect's leg is pulled up and out, it is drawn under this dot. Put a pin under it and lift gently. A pin is much smoother than an insect's leg, but it is likely to pull out a pair of yellow bags fastened to this dot. These bags, No. 4, Fig. 66, are masses of pollen, and if the insect that pulls them out on his leg is more hungry than cautious, he will visit other flowers, and he is likely to leave the pollen bags against a stigma when the same leg gets caught again. A single bag consists of enough pollen grains to fertilize many seeds. Of course the bee may not get the same leg caught again, or the pollen bags may not be broken off. If you catch bees that are visiting milkweed flowers, and with a little patience you can, you are almost sure to find one or more pairs of pollen masses on their legs. Sometimes the bees are too weak to withdraw their legs, and so are caught, and die on the flowers, and very often they accumulate so many of the masses that they are disabled and die; so the beekeepers do not like to have the milkweed in bee-pastures.

Now, with the pollen so rudely snatched from its guests, the ugly milkweed makes the daintiest seeds you can find; so exquisite they are, that it is impossible to sketch them worthily. But you can keep watch of the

Fig. 66. MILKWEED—*Asclepias eriocarpa.*
1. Flower enlarged.   2. Column of anthers and pistils.   3. Column with face of one anther removed.   4. Pair of pollen-masses.

pods. Toward autumn, when the seeds ripen, the pods burst open, and you see at first hundreds of scale-like seeds beautifully arranged along a silken, white centre; each seed rises, expands, dries its wings like a butterfly coming from its cocoon, and finally floats off on probably the softest, downiest ball of fluff in nature.

It is the summer months that the cactus family, too, chooses for flowering; the night-blooming cereus of our gardens for instance, or the strange, weird cacti we bring from the deserts to our public parks, or the prickly pear, or tuna cactus, so common in Southern California on dry hillsides and sandy wastes. We have noticed before how the cactus stores both food and moisture above ground, and how it defends these stores with its dreaded spines and prickles. Some of us have seen the savage cactus hedges planted by the old mission fathers as a defense against hostile Indians.

Naturally, we expect such able plants to produce remarkable flowers; and so they do. The flower of the night-blooming cereus becomes a miracle of beauty, size and fragrance, in order to court the attention of some huge Mexican night moth, and doubtless every other beautiful cactus flower we cultivate, has, in its native land, a story of its own. At any rate, our tuna cactus flowers repay observation. They open only during the brightest hours of the day, deeming it not worth while to proffer hospitality unless there are sure to be crowds of guests. They serve a very thin film of honey deep down in the corolla, but they offer so much pollen that guests seem never to look for honey. Now this clever tuna has a plan for compelling the guests to carry much more pollen than they take for their own use. To discover this trick, you should take a newly-opened flower. The stamens lie back against the petals, but if you touch the filaments, they immediately sweep over to the centre, so that a guest would be buried

Fig. 67. TUNA CACTUS—*Opuntia Lindheimeri*, var. *occidentalis*.
1. Flower before insect's visit.   2. After visit.   3. Branch in springtime showing small leaves.

beneath the anthers. In a few minutes they are back in place ready to bury the next guest; if guests come before the stamens are back in place they must burrow among the anthers, so in any case the guests get thoroughly covered with pollen. Usually there are throngs of guests, especially large woolly bees that need pollen for their babies; and such dusty bees as they are when they leave the flower, perhaps hustled out by newcomers!

Chaparral is a name Californians give to the several kinds of hardy shrubs that form dense thickets on the dry, rocky soil of our hills and mountains. Much of this chaparral chooses summer months for flowering. Some of the shrubby sages of Southern California are still in abundant flower in June and July. Perhaps the plant most commonly called chaparral is the chamisal, or grease-wood. It is a hardy, woody, little shrub, with many small, evergreen leaves that are like spines. It sends out its large plume-like clusters of tiny white flowers throughout the summer months, and as its flowers have no naughty tricks, like the milkweed, they and the bees are of the greatest use to each other. In fact the chamisal is counted in with the "bee-pasture" plants.

One of the Eriogonums, or wild buckwheats, is shrubby enough to be classed as chaparral. You are sure to find it on southern hills. It has more leaves and larger flower clusters than most Eriogonums. The leaves, which grow in clusters, are small, hard, and deep green above; the flowers are first white, then rose-colored, and finally golden brown. This Eriogonum, too, is on the best of terms with the bees. Its flowers furnish honey and pollen, and then trust the bees to cross pollinate them, for, unlike most very small flowers, they cannot pollinate themselves. Find out why.

Now this Eriogonum has a very troublesome plant enemy. The picture, Fig. 68, shows a twig attacked by

Fig. 68. Dodder on *Eriogonum fasciculatum*.

this enemy. The Eriogonum has been able, as you see, to send out only a small stunted flower cluster, but the other plant is in full flower; in fact it seems to consist only of flowers and slender stems. The stems are bright yellow or orange, so the plant is often called gold-thread, but its best-known common name is dodder. Perhaps you have heard it called love-vine, but it must have been an ill-natured person who gave it this name, for it is really a plant of the very worst character. Try to unwind it from its victim, and you will see that the little projections that look like caterpillar's legs are really suckers that have pierced down into the tissue of the host plant, and are absorbing the food that this plant has made, just as the root-hairs of honest plants absorb from the soil their material for food-making. So you can think why this plant has no ordinary roots and leaves and no green color. The dodder attacks many other plants besides this Eriogonum, as you probably know; it sometimes nearly destroys fields of alfalfa.

Now this thieving dodder really belongs to a generally honest family of high rank, to the same family as the morning-glory, as you can see if you compare the parts of the flowers. The little flowers of the dodder make friends with the bees, and mature many tiny seeds, which are scattered in an interesting way. And how do you suppose the dodder seedlings get a start in the world? The seeds germinate on the moist ground, but later than most other seeds, so that there are young plants and new shoots all about them before they sprout at all. The baby plant in the dodder seed, like the older plant, has neither true leaves nor root; it is simply a little stem. One end of the tiny stem, as it leaves the seed coat, glues itself to the soil, but it never sends down roots, nor does it absorb moisture from the soil. The free end goes slowly sweeping round and round, like the tips of morning-glory stems, seeking something to twine about. If it does not succeed before it has used up all the seed food, it must fall to the earth, for it cannot feed itself. But this little, yellow, vagabond baby can do without food for several weeks, and it lies on the ground and waits for some green shoot to grow against it; it seizes this, coils about it, sends down suckers into it, and soon becomes a flourishing plant.

On the shaded slopes of hills, very close to our valley towns, are many bright summer flowers that most people know nothing about, perhaps because they are afraid of summer dust. For instance, within a few minutes' walk from Los Angeles street cars, one can gather in June a few late Mariposas, quantities of rose-colored Godetias, as handsome as Mariposas, brilliant scarlet pinks, the climbing Pentstemon with its vivid, trumpet-shaped flowers, and the first flowers of the scarlet wild fuchsia, with numbers of other, less showy flowers.

The Godetia, pictured in Fig. 69, belongs to the evening primrose family. It has not the appearance of an exclusive flower, but perhaps you have discovered that it keeps its honey beyond the reach of short tongues, and that its style carries the stigma up for pollination by bees or butterflies after its own pollen is shed. The Clarkia is another member of the evening primrose family. It has curious, beautiful flowers in shades of purple, crimson and rose, and you will surely want to study them if the plant grows near you. The Indian pink pictured in Fig. 70, although a member of a humble family, has attained great social success, for it is a favorite of the humming birds. Its petals are not united, but the sepals form a deep, slender tube about them, so that all humble guests are excluded. Can you not read the story of its pollination from the picture? The lower flower is newly opened, and the anthers are shedding pollen; the upper flower is older, the stamens are withered, and the three stigmas are matured.

If you have been watching the Pentstemon pictured in

SOME SUMMER FLOWERS

Fig. 69.
Godetia Bottæ and *Clarkia elegans*.

Fig. 70.
INDIAN PINK—*Silene laciniata*.

Fig. 71, you have learned some interesting things about plants that climb by weaving, as it is called. The flowers, like so many other summer flowers in California, are bright red. Review the story of the pollination of the Pentstemons in Chapter XI, and find out if this Pentstemon has a similar story. Do you think bees can get its honey?

The scarlet wild fuchsia, Fig. 15, Chapter III, is one of the most beautiful of Western wild flowers. From Northern to Southern California, and from the coast to the high mountains, it lavishes its beauty and brightness. In the southern valleys, its leaves are small and gray, and can

Fig. 71. CLIMBING PENTSTEMON—*Pentstemon cordifolius.*

resist the drought; in the north and in the mountains, they are green and more luxuriant; in both cases the foliage is a fitting background for the brilliant flowers. In the Eastern States, this fuchsia is cultivated in gardens, and is called the humming bird's trumpet. You can see for yourselves that it is a humming bird's flower.

Thus far, we have been considering summer flowers, not summer plants, for these plants have been active throughout the rainy season. But California has a goodly number of really summer plants, plants that do not begin their growth from seed or underground stem, till after the rainy season; and so it happens that they have at their disposal a great amount of soil left vacant by the death or dormant state of winter and spring vegetation. For instance, in Southern California hundreds of acres of grain land are in summer taken possession of by such hardy plants as blue-curls, turkey-weed and various kinds of tar-weed. This group of plants we have already studied as "plants that know how to meet hard times," but during the midsummer you may have opportunities to study their flowers more fully. Many of the plants belong to the highest of plant families, the Compositæ, and their "social flowers" have the usual family advantages. There are also plants belonging to smaller families, that have excellent devices for pollination. The blue-curls, pictured in Fig. 15, has, perhaps, the most original trick of all. Pull off one of the corollas, and notice how its slender tube is bent back on itself. It would seem impossible for any guest to get around this troublesome corner; but watch a large bee settle on the lower lip of the flower; the bee's weight straightens the tube and at the same time brings the anthers and stigmas against his back. In younger flowers the anthers are covered with pollen, and in older ones the stigmas are mature.

For those who can spend the summer in our California mountains, there is another world of flowers, for the plants of the higher mountains have wakened from their winter sleep, and will have a good supply of moisture all summer, since there are melting snows, summer showers, springs and streams. In the mountains of Southern California even more than in the valleys, red is a prominent color among summer flowers. As one winds up the mountain trails, at first familiar, scarlet, mountain posies flash out from the greener slopes, the larkspur, painted cup, Indian pink, climbing Pentstemon, wild fuchsia, and, along streams, the scarlet Mimulus. Farther on, along the streams the elegant columbine lifts its scarlet clusters above the bracken; moist slopes are covered with another kind of painted cup, and other kinds of scarlet Pentstemon appear; one kind is a vigorous shrub, other kinds grow in clumps sometimes an acre in extent, and over these brilliant patches the humming birds poise or dart about at their graceful antics; for of course it is the humming birds that make these gorgeous flowers possible.

Of all mountain plants with red flowers, the snow plant is the most unique. Not the flower only, but stem and scale-like leaves are deep crimson. There is no trace of green, for, like many other plants growing in dim forests, this does not attempt to make its own food, but lives on vegetable matter in the soil. It is rarely, if ever, true that these flowers push up through the snow, as some alpine flowers do, but they come early along the edges of the melting snow drifts. This is because the shoots were formed below ground the summer before, and have been kept warm by the snow blanket. The flower clusters last a long time; you may find them in August. Their brilliant color, of course, appeals to the humming birds, but they share the honey with the bees.

The bees have mountain flowers of their own. There are acres of blue gilias for honey, and acres more of blue lupines for pollen. Indeed nearly all the wild flowers you have learned to know in the valley have mountain cousins, many of them in charming mountain dress. There are mountain buttercups, violets, shooting stars, primroses and so on through the list. Besides, there are lovely flowers peculiar to the mountains, such as gentians, saxifrages, and many kinds of lilies. In fact, if you have learned to see and to know the plants about you at home, a mountain journey will give you double the pleasure you would get without an interest in plants.

But you need not go to the mountains to find interesting summer flowers and plants. See how the plants you already know behave in summer. Watch cultivated plants, especially those that climb; whole books have been written on the ways of climbing plants. If you go to the sea, notice how the beach plants protect themselves from the glaring light and the intense heat of the sand; how some of them store water; by what devices they avail themselves of fog and dew; watch their flowers and their guests, and so on. At least, you can watch the wayside weeds, and there is much of interest to learn from them, as we shall see in the next chapter.

CHAPTER XVI.

WEEDS.

What is a weed? An ugly plant, or a troublesome plant, perhaps you will say. But not all of our weeds are ugly. Sometimes, and in some places, our poppy and even one kind of Mariposa lily become troublesome weeds. Some weeds are only cultivated plants relapsed into a natural condition; the wild turnip, which we call mustard, and the wild celery and radish for instance. There are plants considered weeds in California, that are carefully cultivated in older countries; the mustard in many parts of Europe, and the sunflower in Russia, are paying crops. The weedy-looking dock, called canaigre, is very valuable for tanning leather, and we are trying to learn to make use of it. There are still other plants that are treated as weeds when they occur in gardens or orchards, but are encouraged on pasture lands, such as the bur-clover and the filaree.

So the best definition seems to be, "A weed is a plant out of place."

To realize the progress of weeds in California, one must have some idea of the vegetation before the coming of the farmer and the shepherd. John Muir, in a charming chapter on "Bee-Pastures," says: "The great central plain of California during the months of March, April and May, was one smooth, continuous bed of honey bloom, so marvelously rich that in walking from one end of it to the other, a distance of more than four hundred miles, your foot would press about a hundred flowers at every step. Mints, Gilias, Nemophilæ, Castilleias and innumerable Compositæ were so crowded that, had ninety-nine per cent of them been taken away, the plain would still have seemed to any but Californians extravagantly flowery." There are residents of Los Angeles, by no means aged, who remember when the land now occupied by the city of Pasadena and other towns, was in spring time one vivid flower garden, "like a Persian carpet," they sometimes tell us. Even to-day we can see the same thing by going to the foot-hill or so-called desert regions, that are still uncultivated, provided the sheep, "those hoofed locusts," as Muir calls them, have not been before us.

A large part of the valley land to-day is occupied by grain, fruit and garden plants; but there are pasture lands that are practically uncultivated, waste lands in and about towns, and always the untilled waysides. Should we not expect to find our own wild flowers in these places? And do not our native plants dispute with cultivated plants for the possession of the soil, in this case becoming weeds?

As a matter of fact, most California children, even city children, are still able to find quantities of our favorite wild-flowers, but sometimes they must seek them in remote nooks. The spring flowers are rarely common in wastes and along waysides, except in recently settled regions, and it is only here and there throughout the state, that native plants trouble growing crops. Some of these native spring-time weeds are, the poppy, the yellow heliotrope, morning-glory, Calandrinia, owl's clover, sand-lupines, chilicothe, poison oak, bracken fern and cactus. Very possibly you have never seen any of these become troublesome. But in summer time all over the state, after the grain has ripened, there are native plants that take possession of the soil; some of these are the tar-weed, turkey-weed, sunflower, and blue-curls. These plants may be disagreeable and harmful in some ways, but they rarely come early enough to injure the grain.

CALIFORNIA PLANTS IN THEIR HOMES

Really the most troublesome weeds of California are not native plants at all. The plants that occupy the wastes and waysides in towns, and are invading cultivated land everywhere, the malva, filaree, bur-clover, mustard, foxtail, wild oats, shepherd's purse, chickweed, Bermuda grass sow-thistle, pig-weeds, cockle-bur, hoarhound, dock, Spanish needle, fennel, most tumble weeds and their like,— this conquering host is a host of foreigners. Most of them have come here, more or less directly, from Europe, where they have been successful weeds for centuries. It is believed that centuries further back many of these same vigorous weeds were maintaining themselves against cultivation in the older countries of Asia. That is, as civilization has advanced from Eastern to Western countries, these weeds have always pursued cultivated plants. The Indians in the early days of American history, called one common weed, the plantain, the "white man's foot," because it appeared wherever the white settlers went. This particular weed does not thrive in California, but there are weed-travelers that can make themselves at home almost anywhere; the shepherd's purse, chickweed and pimpernel are examples. An English botanist once found shepherd's purse flourishing as a weed on a small island in the Antarctic Ocean. The plants were especially abundant about a sailor's grave, so they probably sprang from a seed that had clung to the spade used in digging the grave.

So it often happens that by the merest accident weeds become colonists in new countries, but they may be said also to have regular routes of travel. Often they come as stowaways in ships, especially in soil used for ballast. Like tramps they steal rides on trains, their seeds hidden in grain, or clinging to imported vegetables, or to live stock. Having once obtained a foothold in a congenial new country, they spread with marvelous rapidity. Burroughs says, "They walk, they fly, they swim; they go underground and they go above, across lots and by the highway. But like other tramps they find it safest by highways. In the fields they are intercepted and cut off, but on the public road every boy, every passing herd of sheep or cows, gives them a lift;" all of which we shall find literally true as we study individual weeds.

Let us begin with our earliest weeds. You remember how quickly after the first rains, the malva, bur-clover and filaree seedlings appeared. The malva was perhaps the earliest, and these plucky little plants are rarely discouraged by the droughts that often follow first rains. In the worst of seasons, they are able to mature an abundance of the little fruits that you call "cheeses," and every section of the cheese contains a seed that is almost sure to germinate. The seeds are small and are blown about far and wide with the wayside dust, or are carried in mud that clings to wheels or feet. Cattle rarely eat the plants if other food is available, so the malva abounds in waste places and is always straying into gardens and orchards.

Bur-clover and filaree cover thousands of acres of waste and pasture land in California. The foliage of the bur-clover is bitter, and stock choose other food first; but if some shoots are eaten the plant quickly throws out others, so it is pretty sure to produce a marvelous number of fruits or burs. In summer time, when the rest of the plant has dried, crumbled and nearly disappeared, the burs remain, sometimes nearly an inch deep on the soil, forming what is called a dry pasture; for so nutritive are the seeds that animals eat the fruits in spite of the prickles, and cattle grow fat on hillsids that look brown and barren. So do sheep if they are allowed to range in these dry pastures, but usually the sheep men avoid them because wool is so much damaged by the burs. Of course the sheep help to distribute the burs that lie in heaps along the waysides. The burs are sure to invade cultivated grounds also; sometimes they germinate in spring and summer in highly cultivated orchards and gardens and become troublesome weeds. In winter, the bur-clover infests lawns and tilled soil generally.

There are two kinds of filaree common on untilled California land; the red-stemmed filaree, or pin clover, which has very finely cut leaves, usually forming flat rosettes, and the musky filaree with coarser, paler leaves that grow nearly upright; the musky odor is more noticeable when the leaves are wilted. The red-stemmed filaree is very widely distributed, and is a good forage plant, but the musky flavor of the other kind seems to be disagreeable to cattle, and they eat it only in limited quantities, so in some places it seems likely to drive out the more valuable kind. Fortunately the red-stemmed kind is the hardier; it can survive severe frosts, and in some soils can even defy the summer drought. We have already studied the fruits of the filaree and can readily understand how it is sure to appear as a weed in cultivated soil.

All three of these foreigners seem to thrive in California even better than at home. There is another immigrant that is vastly more prosperous in this climate than in its native soil, the fox-tail, or barley grass. In England this grass is called the wall-barley, for it is a straggling, insignificant weed that is literally crowded to the walls and crannies by stronger plants; but here it occupies acres of wayside and pasture land, crowding out other and better plants. It is rarely eaten by stock except when it is very young, and its habit of maturing late is also an advantage. It is when other grasses are dying back that the fox-tail pushes its barbed head sunwards, and, unmolested, ripens its seeds. And such a pest as these barbed fruits become! They bore into the nostrils of grazing animals; all summer they lie in wait for anything clothed in cotton or wool, and when once they seize their victim, every movement serves to drive them more firmly in. We would gladly rid ourselves of this most troublesome foreigner, but it seems impossible. The best way to fight it seems to be by encouraging its natural enemy, another weedy but less objectionable grass, known as the soft brome grass. This latter grass has some value as a pasture plant, and so have some other introduced grasses commonly classed as weeds. This is true of the wild oats, which, like the filaree, are so widespread in California that many people refuse to believe them foreigners at all.

In countries of greater rainfall the most troublesome grasses are usually those with perennial stems, which, lying on or just beneath the soil, can root at every joint. We have native grasses of this sort that infest our moist, alkali lands, but the one most troublesome in cultivation is again a foreigner, the Bermuda grass. When once this grass gains a foothold in lawns or other soil kept moist by irrigation, the most constant effort must be made to keep it within bounds.

The mustard is another immigrant whose success should advertise the climate of California. We have already noticed its very rapid and vigorous growth, and its multitudes of seeds. Its prevalence in California means, of course, the loss of thousands of dollars to grain growers every year.

Another foreigner that does mischief in grain and pasture lands is the yellow Melilot, sometimes called bitter clover, No. 1, Fig. 72. It requires a rather moist, loose soil, so it is in the central and northern parts of the state that it is most troublesome in grain fields. In the south it selects choice places, along irrigating ditches and streams, or where there is moist subsoil. It is useless for pasturage, as when it is cut with grain its pronounced flavor taints hay and even affects wheat and flour.

Fig. 72. PASTURE WEEDS.
1. Star Thistle—*Centaurea melitensis.*   2. Yellow Melilot—*Melilotus Indica.*

We must notice, too, the European water cress, which, winter and summer, chokes up our streams. It is interesting to know that in this case, the Western world has retaliated; the American pond-weed, Elodia, is much more troublesome in the water ways of England than the water cress is in ours. The water hyacinth, originally from South America, which is actually ruining navigation in the Florida rivers, is probably not suited to California streams.

There are none of our native plants that, during the growing season, deserve the name of weeds throughout the state; but there are some that are locally troublesome. Sometimes they thrive because of strong underground parts; this is true of the chilicothe, poison oak, bracken fern, morning-glory, yerba mansa, blue-eyed grass, and some kinds of lupines, Umbelliferæ and lilies. Others are rapidly growing annuals that can endure a drought better than cultivated plants, and for this reason become specially troublesome in dry years. The poppy, though a perennial, should be counted here, because it matures so rapidly from the seed. In the drier grain fields of the south in bad seasons, it quite crowds out the grain, but farther north, where the spring bloom is usually prevented by cultivation, it bides its time, and comes up in full bloom in the autumn months. All over the state, wherever it once held sway, it is ready to repossess neglected spots. Other native plants that still assert themselves, even in grain fields, are the yellow forget-me-not, or Amsinckia, the Calandrinia, a Portulaca with bright magenta flowers, the owl's clover, or Orthocarpus, tidy-tips and some small Cruciferæ. All these native weeds seem likely to disappear as our state grows older and the soil is more carefully tilled. Even now they do comparatively little harm, and we are almost sorry to think of the time when there will be no gay flowers mingling with the growing grain.

The weeds of the dry season naturally interfere much

less with cultivation; some must rob the soil of valuable food material, others are certainly disagreeable, but it is possible that some do not deserve the name weed at all. Of our native weeds, sunflowers are the most greedy; they choose the best places, overrunning soil that is loose and rich and not too dry. As soon as the grain is cut, they shoot rapidly upward and soon form thickets miles in extent, affording snug shelter to hosts of sparrows and meadow larks. The myriads of seeds feed the birds, but must take much nitrogenous matter from the soil.

On thousands of acres in California the grain is replaced in summer and autumn by tar-weeds or other resinous, strongly scented plants. The true tar-weeds are Compositæ; they are found in greatest profusion in the central valleys. In the southern counties, the well-known Trichostema, or blue-curls, is sometimes called tar-weed, and there is a very sticky Gilia that deserves its name of skunkweed. The turkey-weed is armed, as you remember, with dense, prickly hairs instead of a resinous coat. It is doubtful whether any of these weeds are particularly injurious to crops, except, of course, such as spring up before the grain is cut; it is even possible that they are useful in helping to loosen the soil and in producing certain chemical changes, but this has not been proved. The odors of some of these plants offend us, and their resinous coating ruins our clothing; along our dusty waysides, too, their dust-laden foliage is not pleasing; on the other hand, some of them have an agreeable, spicy odor, and in masses many are positively attractive to the eye. Some of the tar-weeds have slender graceful stems supporting myriads of starry flowers; the Trichostema is a beautiful plant as long as you do not touch it; and the soft grey-green color of the turkey-weed blends harmoniously with the prevailing summer browns.

The summer weeds that are most troublesome are again the foreigners. One of the worst of these, No. 2,

Fig. 73. WAYSIDE WEEDS.
1. Dock.   2. Knotweed.   3. Sow-thistle.   4. Pigweed.

Fig. 72, is nearly related to the thistle; it is often called the yellow star thistle; another common name for it is the prickly tar-weed, and its Spanish name is tocalote. The plants spring up in winter, but it is not until early summer that they become most troublesome. To the disagreeable qualities of tar-weed they add extremely numerous and prickly flower heads. In pastures they are a great pest, driving out all valuable vegetation. In the central and northern portions of the state, this weed is most abundant and vigorous, maturing sometimes three crops of seeds from May to December. In the south it is usually confined to hillsides, but its place, as a pasture and wayside weed, is fully supplied by an equally objectionable foreigner, the hoarhound. The hoarhound has some virtue as a remedy for colds, but it monopolizes valuable pasture lands, and as a wayside weed is most exasperating; one's clothing, brushing ever so lightly against it, comes away covered with its fruits, enclosed in their little, sharp-pointed, gray calyxes.

Other wayside weeds, the cocklebur, all over the state, and the Spanish needle in the south, have the same unpleasant method of compelling us to distribute their seeds. We have a native "Jimson" weed, which has all the bad habits of its foreign companions. Nettles have still another way of making themselves disagreeable. Other foreign weeds that live along our waysides, and thence invade orchards and gardens, are the sow-thistle, sometimes called milkweed, the pigweed, dock, knotweed, tumbleweed, and the like; humble, homely plants all of them, as their names imply. Yet few of us would complain that nature is so ready to cover neglected places with life and growth. It is not possible to explain fully the hardiness of these weeds; most of them are able, when trodden or cut down, to send out at once vigorous new shoots, and they have various devices for meeting dangers from drought and grazing animals; but their greatest strength probably lies in the wide distribution of their seed. The sow thistle seeds have floaters, and are carried everywhere by the wind. You will see the plants growing on the face of perpendicular banks of clay, and we have to be always watchful to keep them out of our lawns. The other weeds mentioned have, like the mustard, exceedingly numerous small seeds, and in the plant world, as everywhere else, it is often the small things that do the most mischief. In Los Angeles and other towns of Southern California, vacant city lots, and waste places generally, are likely to become small forests of castor-bean or tree tobacco; but it would seem very ungrateful to call these handsome plants weeds.

On the whole, California is not afflicted to an unusual degree with noxious weeds. Many of the worst plant pests of our Eastern States, such as the Canada thistle, burdock, ragweed, chess, and so forth, have not succeeded here, but others, like the mayweed and dandelion, may gain upon us before we know it. Of European weeds, the morning-glory and Russian thistle threaten to become very troublesome. We need to know weeds and to be constantly watchful for those that are likely to become serious pests. The war with weeds must be unceasing. Cultivated plants have lost the power of caring for themselves, and we shall always have to protect them against vigorous weeds that have for centuries fought their own way, some of them all the way from the fertile valleys of Asia across two continents to the western shore of the western world. Work diligently as we may to exterminate the weeds from our own fields, there are always waste places, or perhaps the lands of thriftless neighbors, that serve as nurseries for more of these sturdy waifs.

ERRATA

Page 18; for Hydyodictyon, read Hydrodictyon.
Page 118; for anacetifolia, read tanacetifolia.
Page 207, line 4; for Elodia, read Elodea.

BOOKS OF REFERENCE.

Kerner & Oliver, *Natural History of Plants*, Henry Holt & Co.

This book should head the list, although its cost, $15, bars it from many school libraries. It is beautifully illustrated, is modern, generally reliable and exceedingly interesting, and it is not too technical for the general reader.

Bergen, *Elements of Botany, Pacific Coast Edition*, Ginn & Co.

Spalding, *Introduction to Botany*, D. C. Heath & Co.

Campbell, *Structural and Systematic Botany*, Ginn & Co.

MacDougal, *Experimental Plant Physiology*, Henry Holt & Co.

These are probably the most helpful text books for the general reader or primary teacher on our coast. The first two deal with both the physiology and the structure of plants; both give lists of books for the use of more advanced students. Prof. Campbell's book is especially good for work with the lower plants. There are several excellent laboratory guides, among them one by Prof. Setchell of the State University.

We have no one Flora at all satisfactory for the identification of Pacific Coast plants. One who has mastered technical terms sufficiently to use an artificial key, will find Rattan's *California Flora* the most convenient for work with common flowering plants throughout the state, but

BOOKS OF REFERENCE.

it omits some of the most interesting families. Prof. E. L. Greene's *Flora Franciscana* deals more fully with the plants of Central California. Watson's Botany of California, in two large, expensive volumes, is necessary for anything like thorough work. Other aids are mentioned in Miss Eastwood's Preface to Part II of Bergen's Botany. This Part II, a Key and Flora, is most helpful to one who has some general knowledge of plant families. Miss Parsons' *Wild Flowers of California*, Doxey, San Francisco, is a delightful book and by the aid of its illustrations the names of many of our common plants may be easily found.

For a complete list of the flora of Los Angeles county the following works may with advantage be consulted: *Plants of Los Angeles County*, Anstruther Davidson, M. D., C. M.; *Seedless Plants of Southern California*, Alfred James McClatchie, A. M.; *Lichens*, Dr. Hasse.

John Muir's *Mountains of California* gives most vivid pictures of some features of our flora, and the book should be accessible to all California school children; the well known works of Darwin, Lubbock and Grant Allen are, of course, helpful in the study of the habits of plants, and no one interested in nature study should do without the inspiration of John Burroughs' books.

PRONUNCIATION OF BOTANICAL NAMES USED IN READER.

The accent mark will usually be sufficient to indicate the pronunciation, since it is customary to give the English sounds of the letters in Latin names. If the accented syllable ends in a consonant the vowel is long, otherwise it is short; æ has the sound of long e. A few Spanish names are included in the list, but generally these will be familiar to Californians.

Adenos'toma.
Adian'tum.
Aggrega'tæ.
Al'ga, plural Al'gæ (Al'je).
Alys'sum.
An'giosperms.
Amsinck'ia.
Ar'butus.
Ascle'pias eriocar'pa.
Audiber'tia polystach'ya stachyoi'-
 des.
Bloome'ria.
Brodiæ'a capita'ta.
Calandrin'ia.
Calochor'tus al'bus.
  Catali'næ.
Calyciflo'ræ.
Canāi'gre.
Castille'ia parviflo'ra.
Centaure'a meliten'sis.
Chamisal'.
Chaparral'.
Chilico'the.
Chlo'rophyll.
Clem'atis.

Cni'cus.
Cocome'ta.
Collin'sia.
Compos'itæ.
Conif'eræ.
Cotyle'don.
Crucif'eræ.
Cyc'lamen.
Delphin'ium.
Dicotyle'don.
Dodeca'theron Clevelan'di.
Ellis'ia chrysanthemifo'lia.
Elo'dea.
En'dogen.
En'dosperm.
Equise'tum.
Eremocar'pus setig'era.
Eriog'onum elonga'tum
  fascicula'tum.
Ero'dium cicuta'rium.
Eschschol'tzia.
Euphor'bia.
Ex'ogen.
Fun'gus, plural Fun'gi (Fun'ji).
Gil'ia, commonly Jil'ia, Spanish,
  he'lia.

PRONUNCIATIONS.

Gil'ia dianthoi'des multicaul'is.
Gladi'olus.
Gnapha'lium.
Gode'tia Bot'tæ.
Grevil'lea.
Gym'nosperm
Helian'thus an'nuus.
Hosack'ia.
Hydrodic'tyon.
Legumino'sæ.
Li'chen (Li'ken)
Lo'tus gla'ber.
Lupi'nus sparsiflo'rus.
Macrocys'tis.
Madrone' (madrōn), also madroña and madroño.
Malaco'thrix tenuifo'lia.
Manzani'ta (manzanē'ta.)
Matil'ija (Matil'iha).
Meconop'sis.
Medica'go denticula'ta,
    sati'va.
Megarrhi'za.
Melilo'tus In'dica.
Mentze'lia.
Micram'pelis.
Mim'ulus cardina'lis.
    lu'teus.
    glutino'sus.
Monocotyle'don.
Narcis'sus.
Nastur'tium.
Nemoph'ila auri'ta.
Œnothe'ra cheiranthifo'lia, var. suffrutico'sa.
Opun'tia.

Or'chid (Or'kid)
Orthorcar'pus purpuras'cens.
Pæo'nia.
Papy'rus.
Pentste'mon cordifo'lius.
    heterophyl'lus.
Peuceda'num utricula'tum.
Phace'lia tanacetifo'lia
    Whitla'via.
Piñon' (Pinyōn').
Plagioboth'rys nothoful'vus.
Ploca'mium,
Poinset'tia.
Portula'ca.
Prothal'lium.
Pro'toplasm.
Ranun'culus.
Ri'bes ama'rum.
    glutino'sum.
Ric'inus.
Rhododen'dron.
Sal'via Columba'riæ.
Se'dum.
Sequoi'a.
Sisyrin'chium.
Sola'num Douglas'ii.
Stom'ata.
Tocalo'te.
Trichoste'ma lanceola'tum.
Tu'le.
Umbellif'eræ.
Vio'la peduncula'ta.
Woodwar'dia.
Yuc'ca.
Zauschne'ria.

CALIFORNIA PLANTS

IN THEIR HOMES

BY

ALICE MERRITT DAVIDSON

Formerly Teacher of Botany in the State Normal School,
Los Angeles, California.

SUPPLEMENT

FOR USE OF TEACHERS

1898
B. R. BAUMGARDT & CO.
LOS ANGELES, CAL.

COPYRIGHT, 1898
BY
ALICE MERRITT DAVIDSON

CHAPTER I.

SOME PLANTS THAT LEAD EASY LIVES.

The aim of this chapter is to set children to thinking of plants as living, growing things requiring food and subject to dangers. In the study of the most elementary principles of plant physiology a clear conception of the cell, chlorophyll and protoplasm is helpful, and these ideas are more easily gained from lower than from higher plants. The cell of the water net, even without the microscope, is a definite, tangible object. All green Algæ are excellent for illustrating the giving off of oxygen, and the marine Algæ tell a very striking story of the adaptation of plants to their surroundings. So Algæ have been chosen to illustrate this chapter, in spite of the fact that many teachers are not familiar with them and will naturally be reluctant to use them. But, since in this case neither great knowledge of minute structure nor special skill in manipulation is required, it is earnestly hoped that children will be encouraged to bring material to illustrate this lesson.

Hydrodictyon, or water net, is the most attractive of the fresh water Algæ if a microscope is not available, and it is very abundant, in Southern California at least. As noted in the Reader, the larger nets are usually found much broken up, but smaller entire ones are pretty sure to be entangled with them. The children, in searching for them should use small quantities, examining them in the water, either in white dishes or in thin glass jars in a good light.

For the study of the cell the coarsest pieces of net should be used. The term meshes is applied to the spaces enclosed by the four to seven sides. The teacher should make sure that the children understand that each one of these sides is a cell. The term cell is so misleading that beginners are likely to apply it to the meshes instead. The cells of the water net are usually large and clearly defined, so that with only a hand lens and the pictures, children can gain a tolerably clear notion of their nature, but they will be much more impressed by seeing them under a low power of a compound microscope. They

SUPPLEMENT

will be quick to compare them with familiar objects. A class of little folks with whom I once experimented likened the cells to links of sausage, and the homely comparison showed that they were thinking of the cell as some substance enclosed in a sac ; a truer conception than the botanists who named the cell possessed, for the early botanists, using sections of higher plants, gave their attention only to the cell wall and quite missed the fact that the wall is to the cell only what its shell is to a snail or lobster.

Children should be allowed to find out for themselves that chlorophyll is dissolved by alcohol, and that the jelly-like substance and the granules still remain. As a matter of fact, chlorophyll is confined to small protoplasmic bodies called chloroplasts, but in the water nets these chloroplasts are so closely packed against the cell walls that the cell contents seem uniformly colored. There are other details of cell structure that it would not be wise to impose upon children. For instance, the protoplasm in mature cells lines the walls but does not fill the entire cell cavity, there being much water or cell sap ; in each cell some of the protoplasm is differentiated into nuclei that take part in cell division ; the starch stores are collected about little glistening bodies called pyrenoids, etc. In the *Popular Science Monthly*, September, 1896, the teacher will find a summary of what is at present known about the cell and the division of labor among its parts. Kerner, in the first chapter of his *Natural History of Plants*, gives a vivid conception of the cell. He says : "It is not a mere phrase, but a literal fact, that the protoplasts build their abodes themselves, divide and adapt their interiors according to their requirements, store up necessary supplies within them, and most important of all, provide the wherewithal needful for nutrition, for maintenance and for reproduction."

The bubbles of gas given off by the plants in the sunlight can be easily tested by means of a simple apparatus. Some water net is put under a glass funnel in a glass jar of water, and a test tube filled with water is inverted over the end of a funnel. After the apparatus has been kept in the sunlight two or three days, the water in the top of the tube will be replaced by enough oxygen to be easily tested. The teacher has only to take the tube from the jar and invert it, keeping the open end covered until an assistant has lighted a match or a splinter and blown out the flame after an instant ; when the glowing wood is inserted in the gas, as the cover is removed, the oxygen causes it to burst into flame. It will seem quite credible to the children that this gas which rekindles the flame is the life-sustaining portion of the air. In the first chapter of *Newell's Lessons in Botany* several more experiments are suggested to emphasize the nature of two gases, oxygen and carbonic acid gas. At any rate, the teacher should familiarize the pupils with the names of the gases and some every-day facts about them. For instance, that we are uncomfortable in a close room because the oxygen is being exhausted, so we let in fresh air, which means air with fresh supplies of oxygen; that in all life and growth material is constantly being used up, so that waste substance is being breathed out in the form of carbonic acid gas by all living things; that when material burns more rapidly, wood, coal, coal-oil, illuminating gas, etc., the same gas is given off. The wonderful story of how this product of waste, carbon dioxide, is made over by green plants into food for themselves, and so more or less indirectly becomes food for all the living world, is a story that interests the youngest children.

Children old enough to use the reader are likely to ask questions that will demand knowledge of some of the following facts:— the water in which Algæ live always contains carbon dioxide or CO_2—a compound of two parts oxygen and one of carbon—which it has absorbed from the air. The formula for water is H_2O—two parts of hydrogen to one of oxygen. Now while starch and similar substances, all classed as carbohydrates, are known to consist of carbon, hydrogen and oxygen in definite proportions, it is not possible to artificially manufacture them from these elements. It is only in the cell laboratories of green plants that this combining or synthesis (photosynthesis some botanists call it) takes place, so green plants only can bridge the gulf between the inorganic and the organic world. Just how this is done is not known. The protoplasm of the chloroplasts must be the builder, the sun's rays supply the motive force, and there are theories about the part played by the chlorophyll, but no one claims to have solved fully this mystery of life.

These primary organic substances are, by chemical changes, converted into all the other compounds, such as albumen, cellulose, starch, fats, pigments, etc., of which the bodies of plants and animals are composed. The technical term applied to these changes in plants is metabolism. The impelling force for these chemical changes is obtained by oxidation; that is, the protoplasm withdraws oxygen from the air and uses it in burning a portion of the carbohydrates, but only a fraction—sometimes $\frac{1}{25}$ or $\frac{1}{30}$ of the amount manufactured by the plants. So plants, like animals, respire, that is, use oxygen and give off carbon dioxide. Other inorganic substances than carbon dioxide and water are used in these chemical processes. Nitrogen is necessary and so are several mineral salts, such as sulphur and iron.

SUPPLEMENT

Some marine plants use much iodine, or soda, or calcium, each plant selecting from the thousands of gallons of water flowing over it the special mineral it requires.

These manufactured organic substances of course serve primarily for the growth of the plant, that is, for the extension of protoplasm and the envelopes produced by it, but surplus food is stored, often in the form of starch as in the water net. The starch can be identified by putting iodine on the colorless plants taken from the alcohol, and comparing the color of the grains with color of wheat flour similarly treated. The iodine of commerce will answer.

Beyond comparing the size of the smallest nets with the largest mature cells, the children will have to take the story of reproduction on authority, but it can be made an interesting one. Develop from the children the fact that the little protoplasts join hands and form nets for safety. They are not so easily swept away or swallowed by little animals. The statement that the stronger method of reproduction by resting spores occurs as danger approaches is based on laboratory experiments, in which the two kinds of reproduction are induced by simulating favorable and unfavorable natural conditions. The spore formed by the union of the two tiny protoplasts does not grow directly into a new net. It first divides into several peculiarly shaped cells which may ultimately become nets.

Instead of water net, the children may find Algæ that consist entirely of threads or filaments. They may be rough to the touch and branched, as No. 5, Fig. 4, which represents Cladophora, or water flannel. No. 6 is an outline drawing of part of the same under the microscope. The individual cells closely resemble those of water net. Or the children may find pond-scum or water silk consisting of slimy unbranched filaments. Nos. 3 and 4 represent two kinds of water silk, Spirogyra and Zygnema, under the microscope. As the chloroplasts of these plants are spiral bands or more or less star-like masses, these plants are particularly attractive under the microscope. All of these filaments lengthen by a process called cell division. From about the middle of the cylindrical cell wall a ring of cellulose, as the substance comprising the cell wall is called, pushes its way toward the center of the cell until it becomes a circular partition dividing the former cell into two cells which soon attain full growth. In Cladophora branches push out at first like little swellings. All of these filamentous Algæ reproduce by fragments, just as higher plants are propagated by slips or cuttings. Cladophora reproduces also by zoöspores, that is, the many protoplasts into which a mature cell may divide, escape and swim about for a time before growing into new

CHAPTER I

filaments. The term spore is applied to a protoplast that is capable of growing into a new plant, and the prefix "zoö" means behaving like an animal, that is, moving. The pond-scums do not produce zoöspores but form strong resting spores by a union of the contents of two cells, as shown in Fig. 4, Nos. 1 and 4. These conjugating filaments are frequently found in nearly exhausted pools; without the microscope they look rather broken up and spoiled.

Among the plants brought by children to illustrate this lesson there may be a dark blue-green slime that they may have found coating the sides of ditches or reservoirs. This is likely to be Oscillatoria, and it belongs to the lowest group of plants, the Protophytes, the group that includes bacteria. A minute bit of this slime, undisturbed in a dish of water, will show radiation of the filaments in less than an hour, and in twenty-four hours it will exhibit in a striking way the movements of the plants in order to secure better conditions for food. To be sure, this movement may be mistaken for growth. The children will be quite likely so say that the plants have "sprouted" or floated. The microscope will demonstrate the movement, which is mainly an oscillating one. But in default of the microscope, children will take the fact of movement on authority and will be interested in watching evidence of it and in thinking out its use to the plants. They will be quick to discover the slime on damp walls and flower-pots, and green film on stagnant pools; and they will gradually become impressed with the fact of the existence of a wonderful microscopic world. This world appeals to children's imaginations, and while most unicellular plants, especially bacteria, are much too small to be handled in elementary work, a well informed teacher will find no difficulty in impressing children with some practical truths about them.

If microscopes are used at all, some unicellular organisms are pretty sure to be encountered incidentally in work with fresh water Algæ; the pretty, green, crescent-shaped desmids, perhaps; or little active, transparent infusorians; or, most common of all, little, brown, boat-shaped organisms, diatoms, that move about in a jerky way. The desmids are plants, the infusorians animals; the diatoms are still included among plants in most text books in Botany, but excellent authorities have of late relegated them to the animal kingdom; so thus the fact that there is no sharp dividing line between the plant and the animal world is illustrated.

Children who live by the sea, or even those who visit the sea only occasionally, should not miss the pleasure of becoming acquainted with some common marine Algæ. The teacher can, in one visit to the sea, lay in a stock of specimens that will last for years. The coarse, large kelps will stand repeated drying and soaking, and the more delicate plants can be preserved in all their attractiveness. On the rocks at low tide they find the bright green ruffled and translucent little fronds of sea lettuce (Ulva), and perhaps some of the filamentous green Algæ. The common marine Cladophora has much shorter filaments that the fresh water plant and grows in spongy tufts that are well able to resist the waves.

Rock weed (Fucus) is the most abundant of the Algæ in shallow water. The plant is rarely more than a foot in length, and it has many flattened branches that bifurcate repeatedly; in the ends of the branches little dots are apparent. The rock weed is not especially attractive, but it grows so near the shore that at low tide it can always be studied on its native rocks, and the tenacity of the holdfast, and the toughness, elasticity and flexibility of its branches tested. Entire plants of Macrocystis are frequently thrown up on the beaches with holdfasts a foot or more in diameter. There are likely to be fragments of many other kelps with interesting holdfasts and air sacs. One kind common on Southern California beaches is the Neyreocystis. The stem is unbranched for many yards and terminates in a great air sac six inches or more in diameter, which floats a whorl of branches with leaf-like expansions. A particularly graceful brown Algæ, the sea oak, can be found at the lowest tides growing on the rocks. It has a solid holdfast and the stem soon branches into divisions that resemble lobed oak leaves. Farther up the divisions become a series of air sacs that resemble strings of amber beads. In order to appreciate fully the significance of the floaters, one must keep in mind the fact that sunlight as well as chlorophyll is necessary to plants that make their food from inorganic matter.

Seven hundred feet below the surface of the sea there is absolute darkness, and the zone of vegetation is usually limited to within one hundred feet of the surface. This latter fact is due not merely to the dimness of light. It is easily shown by experiments that, of the rays of different colors and wave lengths that together make up colorless daylight, only the red, yellow and orange rays help in manufacturing the primary organic material. Other rays have the reverse effect, breaking up this material for further changes. Now salt water absorbs the red, yellow and orange rays, that is, it is blue; so it is well for the larger brown Algæ to be held near the surface by the air sacs.

Our red Algæ are much smaller, and those that grow in deep water cannot reach the surface, hence the necessity for the red pigment; for this pigment has the quality of florescence, that is, of absorbing some rays and giving out others. In this case it changes blue rays to some extent to red, yellow and orange ones, and so enables the chloroplasts to perform their mission.

Red Algæ that grow near the shores are not usually brightly colored, in fact they are almost sure to be mistaken for green Algæ by the novice. The genus Gigartina illustrates this fact well. Gigartinas are very common on our coast. They are characterized by little elevations, some of which bear globular spore fruits. One species, *Gigartina horrida*, has very narrow divisions so that the plant resembles bunches of coarse, dark green chenille. Another species has fronds so wide that the children call it apron kelp. This grows in deeper water and is a deep purplish crimson. There are intermediate species that show all possible graduations of color, according to the depth of water. As these Gigartinas bleach on the beach they present a great variety of color, the red or purple color fading to pink, and finally vanishing and exposing the chlorophyll, which in turn fades out. This fact has given rise to the term "calico" kelps that children often apply to them. Our most common red, feather moss, Plocamium, and many other delicate and brilliantly colored species grow in deeper water. Another common genus among red Algæ is Corallina, named from the resemblance to coral. And here again the idea of the similarity in form of lower plants and animals can be impressed, for coral is animal, but Corallina is plant, while a very pretty, graceful thing that they will call brown feather moss, seems much more like a plant than does Corallina, but it is really animal, a hydroid. The Corallinas are small jointed plants and are very brittle when dried, because of the amount of lime they contain. Like the Gigartinas, these plants show the presence of chlorophyll as the red pigment fades. The use of the joints to the Corallina and the fine dissections to the Plocamium and other red sea moss can be thought out.

With the microscope many interesting facts of cellular structure can be shown. The filamentous Algæ consists of rows of cells; the very thin fronds have their cells arranged in but one or two layers. A section through some of the thicker fronds, like the rock weed, shows cells irregularly netted or interwoven, compactly at the outside but very loosely within, a structure which is obviously adapted to the environment of the plants. If it seems best to the teacher, some facts of reproduction may be taught, but they must, in elementary work, be taken mainly on authority. The marine green Algæ reproduce by zoöspores. Many brown Algæ, and all red Algæ also reproduce assexually, i. e., by simple cell division into spores, either zoöspores or non-motile spores. But some brown Algæ, rock weed and sea oak for

SUPPLEMENT

instance, have a higher method of reproduction. Within the conceptacles, visible as dots to the unaided eye, are antheridia containing minute fertilizing or sperm cells, and oögonia containing much larger egg cells. A sperm cell and an egg cell unite to form a strong spore. Red Algæ have also a higher form of reproduction, As the result of the union of two unlike cells there grow spore fruits. These spore fruits are conspicuous on the Gigartinas, and are frequently found on other red Algæ. The growth of new Algæ from fragments is mentioned in the Reader.

CHAPTER II.

HOW SOME PLANTS BEGIN LIFE.

The first aim of this chapter is to develop some fundamental facts in the physiology of higher plants. There are also several minor reasons for taking up work with seedlings in the autumn. There is more time for it, since there is less out of door plant life clamoring for attention; the temperature is favorable, and children can be prepared to enjoy the native seedlings that the rains bring. The seeds do best planted in sand in wooden boxes. The sand should be kept moist but not wet.

An idea to be emphasized from the first is that the seed is a little plant with its equipment of food for beginning life by itself. The castor bean seems to me to illustrate this idea more forcibly than any of the other common seeds. The ideas of seed protection and seed distribution are made only incidental here; they should be impressed throughout the year, but the early summer affords the most striking illustrations. The poisonous quality of the castor bean is well established by medical records. The seeds are sometimes fatal to man, and I have not been able to find that any animals eat them. The violent expulsion of the seeds is likely to occur on sunny days following a dense fog or a rain. Mature seed vessels, moistened and then kept is a sunny place in the school room, are pretty sure to throw some of their seeds, but even out of doors many of the seeds are dismissed gently and fall near the parent plant.

The structure of the seed is most easily taught by beginning with germinating seeds and then comparing them with unsprouted ones. It is best to keep the seeds in water about twenty-four hours before dissecting them, and the hard seed coat should be cracked with a hammer. The castor bean has also a distinct inner seed coat, but this is hardly worth noticing, particularly as it is not present in all the other seeds. An ovule before fertilization has an opening to the embryo sac through which the pollen grains enter. This orifice is called the micropyle, and through it the radicle forces its way even

SUPPLEMENT

when it has become closed by the seed coats. The knob at one end of the castor bean is the thickened edge of the micropyle and it becomes an excellent absorbing organ.

It may seem petty to distinguish between the root and stem in the embryo, but it is well to note from the first that leaves are borne by stems, not by roots, and that roots grow from stems, not stems from roots. The technical terms for seed leaves, stem and root, have been omitted until later on in the chapter in order that the real nature of the parts be not obscured. In fact it is only in deference to custom that they have been introduced at all in such an elementary work. The castor bean seedling early gives off side roots at the end of the stem, usually in four groups, each group from a pair of woody fibres, as can be seen from a cross section of the stem. The fact that the growing cotyledons absorb the food, both for themselves and for the rest of the embryo, seems to me to be clearly illustrated by the natural development of the seedling, but the endosperm may be removed from one of two seedlings of equal size and the result noted; or this experiment may be performed with any of the other seedlings. Obviously some of the stored food is in the form of oil, and unless the teacher takes special care, this fact will be recorded in grease spots on the floor. The oil of the castor bean, like starch that forms the bulk of corn, wheat and many other seeds, is not nitrogenous, but all seeds have also some albuminous or proteid substance. This is necessary since the living and growing parts of all plants as well as animals contain much proteid material. The endosperm of the castor bean is, in reality, as well as in appearance, similar to cheese, but it is hardly practicable to test proteid substances in elementary work. Oily seeds do not decay so readily as starchy ones, but they germinate more slowly. Still the castor bean gets well above ground in three or four weeks. Some observing pupil may notice the glands at the base of the castor bean cotyledons. These are very marked in the mature plant and will be referred to later on.

Pine seeds are introduced mainly because of the attractiveness of both cones and seedlings. Seeds of the Monterey pine (Pinus insignis) can be had at seed stores. They germinate almost as rapidly as the castor bean. Piñons are difficult to grow and require much time. Cone-bearers will be taken up again in Chapters VIII and XIII. They constitute an interesting family and have a marked individuality. John Burroughs brings out this idea charmingly in "A Spray of Pine," in "Signs and Seasons."

Children may be led to trace the other seeds suggested back to their plant homes and to note their devices for protection and distribution, but cultivated plants, that is, plants that have been propagated artificially, are usually not so well adapted to care for themselves as native plants or hardy weeds. On the other hand, plants that are prized because of the food stored in their seeds grow rapidly; so the time-honored bean or pea and corn are not likely to be supplanted for school room uses. The morning glory is suggested because the embryo has cotyledons even more leaf-like than the castor bean; but the seed is so small that the seedlings are slow to develop the next leaves. The difference between seeds with and without endosperm is by no means a fundamental one. The common bean and the squash cotyledons have simply absorbed the seed food before entering on the resting stage; the cotyledon of the castor bean and corn absorb it during germination.

Nasturtium seeds are very easily grown in the school room in our climate. Like the pea, they keep their seed leaves underground, though they do not seem to need this protection so much since their pungent taste probably protects them from the attacks of animals. The wrinkled coat of the nasturtium seed helps to anchor it so that the radicle more easily penetrates the soil. The glutinous envelope of squash seeds serves the same purpose; so do the hooks of the bur clover, for the seeds usually remain in the bur. Acorn and walnut seedlings always appeal to children. The uses of the cotyledons of the acorn should be developed. They protect the rest of the embryo and liberate it from the shell, besides supplying it with food. The walnut cotyledons contain much oil and the seeds are well protected. On the tree the "shells" or ovary walls are an inconspicuous green color and are extremely bitter; later, on the ground the shells are again the color of their environment and are still troublesome to marauders.

The corn seedling presents some difficulties, but it is so convenient for use in experiments that it is introduced here. A considerable quantity should be grown for use in connection with the next chapter. Children need not be told that the kernel is ovary as well as seed, and that some botanists call the absorptive organ the scutellum. The main point is that they understand the functions of the parts. As children become interested in growing seedlings they are likely to make interesting discoveries. They may find barley heads with every seed germinating in its husk, or seeds germinating inside lemons or squashes, and they will see that the lemon seeds have several embryos within one seed coat. They will soon learn to recognize common out-of-door seedlings like the filaree, malva, bur-clover, grasses, mustard, nettles, geraniums, marigolds, etc. The little arrow-head

SUPPLEMENT

cotyledons of the malva along our waysides may be as welcome as heralds of our green season as is the first robin's note to our Eastern friends. Lesson V in the Reader assumes that the teacher will encourage pupils to observations of this sort. It is specially urged that much attention be paid to plants encountered every day, to wayside plants like the filaree and bur-clover. The very fact that these two introduced plants have gained such a foothold in our state proves that they have many traits worth studying. But the great point to be gained, is the feeling that there is a world of interest in common things.

Much of the physiology of seedlings must be taken on authority in elementary school work. But there are many simple experiments that require little time or apparatus. If scales or balances are available, the amount of water seeds absorb in a given time can be tested, or the teacher can devise other means for measuring the water used. Many seeds germinate readily in moist air. The development of these seedlings can be very clearly seen; they show root-hairs and root-tips beautifully and lend themselves to many experiments. A moist air chamber is made by simply putting pieces of wet blotting paper in any convenient dish and keeping it rather closely covered. A piece of glass is a convenient cover because the seedlings can be watched without letting the moist air escape. The seeds should be soaked, then simply laid on top of the paper. They germinate better in the dark as can be shown by keeping one of two dishes covered with a piece of black cloth. In the experiment of determining where the growth of the root takes place, common pen and ink may be used. Notice that no matter how the seeds are placed on the paper the roots always point downward. To show that this is not due solely to the moist substratum, pin seedlings to a cork after the radicle has first broken through, and keep the apparatus in a moist air chamber. Bell jars or any wide-mouthed glass vessels are very convenient for use in experiments with seedlings. Wheat grows prettily on a sponge placed over a glass of water when the whole is covered so as to keep the air moist. Mustard seeds grow very quickly on a piece of gauze stretched over a glass full of water, the whole covered with a large glass vessel. If one side of the apparatus is kept light and the other darkened, the stems grow toward the light and the roots away from it in a very striking way.

Any seedlings enclosed in a fruit jar will illustrate transpiration. To make sure that the moisture is given off through the epidermis of the leaves and stems, the cut ends of the stems should be sealed with wax or gum of some sort. Corn leaves transpire so rapidly that the moisture from them appears on the sides of the jar in a few minutes. A piece of polished marble in the bottom of a box of growing seedlings will become etched by the acid secretion from the root-hairs. Probably these simple experiments with the deductions to be drawn from them will occupy as much time as the plant lessons are entitled to in primary schools. The teacher who can find time for more will find them suggested in the more recent books mentioned in the preface, particularly in Bergen and Oels.

As soon as seeds absorb moisture the insoluble stored food is changed to soluble forms, and other chemical changes that always accompany growth and nutrition occur. As stated in the supplement to the first chapter, these chemical changes require oxidation, that is, the plant respires. Higher plants have not, like higher animals, special organs for respiration, but the process is going on over the whole of the plant throughout its existence. Whether the respiration of plants should be taught in the grammar grades seems to me very questionable. The experiments bearing on this subject require some knowledge of chemistry and are not likely to be comprehended. Respiration, too, is to a considerable extent the reverse of photosynthesis, and if there is an attempt to teach both processes the whole subject becomes confusing.

Of the experiments suggested above, by far the most impressive one is growing corn in moist air. The root-tips and root-hairs are beautifully distinct. The teacher has only to remember that the root-hairs are so delicate that they shrink after a few minutes of exposure to dry air. The relation of the number of hairs to the plant's facilities for obtaining moisture is very interesting. The mustard seeds germinating on gauze, and the wheat on the sponge, produce very striking root-hairs so long as they are in contact with moist air only, but when the roots have reached the water in the glass they cease entirely to develop hairs; that is, root-hairs are abundant so long as moisture is difficult to obtain, but they are obviously unnecessary to roots immersed in water. By actual count there are from ten to four hundred root-hairs per square millimeter. Root-hairs on plants grown in the soil become very firmly united with it, so that it is impossible to entirely shake off, or even wash off, adherent particles. This adhesion is due partially to the solvent acid of the hairs, but it is also because of the avidity of their cell contents for the molecules of water adhering to molecules of the soil. The absorbent cells must be able to extract considerable water from soil that appears dry. Root-hairs perish on the older parts of roots, but there is always a zone of absorbent cells just back of the tip.

SUPPLEMENT

The teacher must make sure that the root marking experiment is thoroughly comprehended. The marks are at first equidistant, but the mark on the tip is carried further and further on, while the others remain exactly as they were. If roots were to grow throughout their length through the unyielding soil they would necessarily become much distorted. The root-tip is protected by a little cap that is renewed as fast as it is worn away by friction. The remarkable properties of root-tips are treated at length by Darwin in his "Power of Movement in Plants." In his enthusiasm over the results of his experiments with seedlings he says: "It is hardly an exaggeration to say that the tip of the radicle, thus endowed, and having the power to direct the movements of the adjoining parts, acts like the brain of one of the lower animals; the brain being seated within the anterior end of the body, receiving impressions from the sense-organs, and directing the several movements." Some German critics consider this an extravagant statement, but one of them, Sachs, in his "Physiology of Plants," explicitly states that the end of the root is sensitive to pressure and to moist surfaces, and that as a result the root curves so as to avoid the obstacle and reach the moist substratum.

The tendency of the main root to grow directly downward can be easily shown by pinning seedlings to cork in the moist air chamber so that light and moisture are everywhere equal. The roots of seedlings grown on gauze, or on a sponge over water, clearly grow away from the light. Experiments to show the spiral movement and sensitiveness to moisture and pressure require more time and apparatus.

The padding of roots, which protects them against evaporation and mechanical injury, can also be noted here. The natural length of roots will not be very accurately determined from seedlings grown in shallow dishes. Wheat, out of doors, has been known to send down roots to a depth of seven feet, and the sum total of the length of all the roots of an oat plant is sometimes as much as one hundred and fifty feet. Seeds grown in a crowded condition show the power of roots to bind soil together so that it will resist the force of wind or rain. But the main point here is to emphasize the great extent of the absorbing surface furnished by this elaborate root system, that is, its capacity for taking in food.

Some of the minerals needed by all plants are potassium, sodium, calcium, sulphur, magnesium, phosphorus and iron. The proportions of the minerals required vary with different plants as every agriculturist knows, but by striking an average, these salts, with necessary gases, can be artificially combined into a nutritive solution in which plants will grow to maturity, whereas seedlings grown in pure water perish soon after the food in the seed is exhausted. Further experiments show that the presence of iron is necessary to the formation of chlorophyll; and it is well known that nitrogen is a constituent of protoplasm. Now, although nitrogen forms a large percentage of the air, experiments seem to prove that it is of service to plants only when it enters in combination with other substances. So the nitrogen required by plants is absorbed mainly from the ground, though sometimes from the air, and it is in the form of nitrates or of compounds of ammonia. Nitric acid and ammonia arise from the decay and oxidation of dead organic matter, and since all putrefaction is due to bacteria, we may say that nitrogen is prepared for the plants by bacteria in the soil.

Corn and castor bean seedlings are excellent for illustrating the ascent of crude sap, but their own red pigment should be located before the red ink is used. The natural pigment is in or near the epidermal cells, where it is supposed by some botanists to serve as a screen; the artificially colored fluid will always travel by way of the woody tissue. Of course for higher work more accurate experiments than this have been devised for showing this function of woody tissue. The fact of the existence of this woody system, characteristic of higher plants, is perfectly apparent in seedlings, but its full significance and the reasons why lower plants can do without it, need not be discussed here.

The cellular structure of stems is far too complex a subject to be treated fully in an elementary work, but it will be again referred to in Chapter VIII. Drawing No. 2, Fig. 10, which is of course partly diagramatic, is introduced here only to emphasize the division of labor among different kinds of cells. This conductive tissue of wood cells and vessels extends to the farthest tips of the leaves. The epidermis of the leaves serves for protection and prevents excessive evaporation. It has a cuticle that is almost impervious to water. The moisture that is given off by leaves escapes mainly through the pores or stomata, and the two guard cells of the stomata, by separating or closing, regulate the amount of transpiration. The typical horizontal leaf has, within the epidermis, palisade cells on the upper side and spongy tissue beneath; that is, the upper cells are little cylinders compactly arranged and at right angles with the epidermis; below, the cells are so loosely arranged that there are many air passages that communicate with the outside world through the stomata. These intercellular canals also convey the moisture that is evaporated from the cells.

SUPPLEMENT

And now we must consider the problem of how the current of water with its freight is raised from the root-hairs to these cells in the leaves, and the more one looks into investigations on this subject and reads the various conclusions and theories of leading botanists, the less is the inclination to offer any solution of the problem to children. The old theory of capillarity seems to be quite abandoned. The laws of the diffusion of fluids and gases separated by a membrane, such as a cell wall, perhaps explain the great avidity of cells for water, especially of cells containing stored food and active protoplasm. At any rate the water of the soil is drawn in by the root-hairs and forced on by contiguous cells, and the force is named root pressure. Root pressure is supposed to account for the rise of sap in maple trees in the spring, the "weeping" of cut grape vines, the drops of moisture often apparent on corn or wheat seedlings, and the gush of sap from the cut leaves of the agave and other plants of arid regions. This force is most active when stored food is being rapidly used in new growth. Apparatus has been devised for measuring this force, and the results can be found in any text book that treats to any extent of plant physiology.

There seems to be no doubt that the evaporation from the leaves is a great lifting power. That is, the water given off by evaporating cells near the surface is replaced by diffusion from adjacent cells, these in turn draw on cells below, and so on. In this way an ascending transpiration current arises. It seems to be quite proven that the ascending current is mainly through the walls of the wood cells and vessels, rather than through their cavities; and theories have been advanced to meet this and other phenomena. After all, the main point is that nutritive salts and nitrogen provided in the soil do reach the laboratory cells in the leaves where they can be combined with the organic material manufactured there.

This manufacturing of organic matter from carbon dioxide and water, or photo-synthesis, has perhaps been already sufficiently emphasized. As stated in the Reader, carbonic acid gas is absorbed directly from the air; in the epidermal cells it has become carbonic acid and the acid is absorbed and decomposed mainly by the protoplasm of the palisade cells. The oxygen is of course given off through cell walls into intercellular passages and thence through the stomata to the outside world. The first product of photo-synthesis is not well understood, and since in most plants it becomes visible first in the form of starch in the chloroplasts, many teachers think best in elementary work to simply call the product starch and the process starch-making.

CHAPTER II

The subject of metabolism, that is, of the farther transformations of primary organic substances, and the fact of respiration, have been referred to already. Oels' "Experimental Plant Physiology" suggests many experiments under this head. Kerner, in Vol. I, Part 2, of his "Natural History of Plants," classifies these products as follows:—under building materials as albumens (albumen, casein fibrin, etc.), cellulose (changed sometimes to lignin or cork) and starch (a mixture of cellulose and granulose); and under accessory substances, pigments, including chlorophyll, sweet-tasting substances, oils, resins, balsams, fats, alkaloids (nicotine, quinine, etc.), glucosides (saponin, tannin, etc.), organic acids, organic salts, amides and ferments. But the subject seems hardly suitable for children. They can of course think of various plant products, and can reason out the fact that the building materials originating in the leaves must be transported to any part of the plant requiring food or acting as a store-house. So it becomes apparent that, besides the ascending current of raw material, there is also a stream carrying organic food, whose direction is in the main downward. The movement of this stream is explained by the law of diffusion. As any cell uses up material, the same material will be at once replaced from a contiguous cell, which cell repairs losses from the next cell and so on. The sets of cells through which this distributing current passes will be referred to in the supplement to Chapter VIII, when the subject of the division of labor among the cells will be further emphasized. The fact that metabolism and growth go on even better in darkness seems to be thoroughly established by means of experiments. It is found that the greatest daily growth is usually just before sunrise.

SUPPLEMENT

CHAPTER III.

PLANTS THAT KNOW HOW TO MEET HARD TIMES.

The aim of this chapter is to develop the adaptation of plants to a warm, dry climate. Another chapter will deal with plants that spring up under reverse conditions, that is, with plenty of moisture but a limited amount of heat. As some knowledge of the relations between climate and vegetation underlies all intelligent study of geography, this feature of plant study seems especially important.

The first point to be emphasized is the paramount necessity for water in plant economy. Water itself is a building material; it enters into all substances that compose a plant; it is also necessary as a solvent for other building material; besides this, the transpiration current is the main propelling power. If children have not been set to thinking of water in the soil and of evaporation in connection with their geography, this should be done now. They usually know that as we dig down we come to moist earth. Perhaps they know that in California stream beds there is often water below the surface. In our narrow cañons and on steep north slopes moisture is usually evident. It is easy to show the relations of heat and dry air to evaporation by some common illustration, such as the drying of clothes.

It will be well if the teacher can supervise the collecting of these autumn plants. Happily the plan of spending an occasional half-day in out-of-door study is gaining favor in our schools. In this case half an hour in a vacant lot or field and along the roadside would give an impulse to the collection, but if a visit to a stream-bed or a cañon be feasible the lesson will be much more impressive. In the suburbs of Los Angeles, along the Arroyo Seco, all of the plants referred to in the Reader, and many others, could be found in a half-mile walk. In other regions the flora will be somewhat different, but types of dry climate vegetation will be abundant. If there are no native plants of the fleshy type, use the cultivated Sedums (hen-and-chickens, live-forever) or the fleshy plants often used to cover our grounds and parks because they are better suited to our climate than grass. One prominent group of ever-present plants, the cosmopolitan weeds that flourish on the outskirts of cultivation, the pig-weed, cockle-bur, knot-grass, sow-thistle, Spanish-needle, etc., have received little notice here, because a chapter is given to weeds later on. But this is a good time to call attention to them and to see that their names are known, so that in the course of the year the children will become impressed with their prevalence. Of course it is not necessary to know the botanical names of all the plants handled, but if the teacher succeeds at all in fostering children's interest in plants they will want many names, and the teacher should be able to give the names of the native trees of the vicinity, and of the most common weeds and most attractive shrubs and flowers—common names when they are well established, otherwise the botanical generic name.

In the field work, the dead annuals should be pointed out, and there will usually be some woody perennial herbs that have partly died back. The perennials that perish to the ground will be noted in the next chapter. If a stream is visited, water-cress may be found; note the first fact that the pungent taste probably does not recommend the plant to animals as it does to us. Sometimes very bright summer flowers are found along our water-ways, the yellow *Mimulus luteus* with its brown spotted throat—the children may call it snap-dragon —or the bright red *Mimulus cardinalis*, or the bur marigold, (*Bidens chrysanthemoides*) a large, showy, yellow Composite. If the stream is in a narrow cañon there will be brakes and horsetails, and perhaps our stately Woodwardia fern. But these plants do not really belong to the type we are studying; they are water-loving plants and should be noted here only on account of their individual merits and by way of contrast. The wild grape, too, grows only where it has plenty of water and so can afford to expose its leaves as fully as possible to the light. It forms a beautiful illustration of leaf-mosaic, but this topic belongs to the next chapter. Of course the clematis fruits, with their beautiful device for seed distribution, would not be passed without notice.

Perhaps the poison oak and nightshade (Solanum) are hardy enough to be classed with the other plants, but they, the poison oak especially, seek shaded places. They have very long roots with thickened parts that store moisture, and although our nightshades unlike the European *S. nigraum* are not poisonous to man, and some grazing animals can eat poison oak; on the whole these plants seem to be little molested by the animal world. The poison oak begins its period of rest in autumn, but *Solanum Douglasii*, about Los Angeles

SUPPLEMENT

at any rate, flowers without any intermission. The California holly (*Heteromeles arbutifolia*) has leaves typical of a dry climate.

In treating the topic of the protection of plants against the attacks of animals, it is easy to make too sweeping statements. Plants that are poisonous to some animals are eaten with impunity by others, and plants sometimes have an evil reputation in this respect that they do not deserve. But almost any locality furnishes some illustrations of this device for protection. In many parts of the state the poisonous *Datura meteloides*, sometimes called "Jimson weed," is common in summer and autumn. The point in teaching this topic is to arouse children's interest in the subject so that they will compare observations, and make inquiries about the supposed poisonous plants in their own neighborhood. A taste and odor that is disagreeable to us may not be equally so to grazing animals or caterpillars. Cows often eat bitter herbs that make their milk utterly unpalatable to us. On the other hand, plants that seem to us quite inoffensive, ferns for instance, seem to be never molested by animals. Hunger often leads animals to devour plants that are so rough and woody as to appear quite safe; but some of our California dry-season plants seem to have rendered themselves absolutely inedible by these means. We must remember, too, that plants do not need to be absolutely invulnerable to attack. If they can protect themselves against their most common foes they stand a good chance for survival. The turkey-weed (*Eremocarpus setigera*) is a plant that sheep refuse to eat under the direst necessity, a fact that goes far toward accounting for its prevalence in many parts of California, for sheep have made sad havoc with our native flora. The cactus will be taken up later on, when, it is hoped, the children will have acquired the habit of alertness in discovering plants' devices for protection against animals.

Plants have various methods of protecting chlorophyll against too much light. In unscreened leaves the chloroplasts group themselves against the sides of the cells in order to avoid intense light. This can be illustrated by fastening dark paper over part of such a leaf on a very bright day; the unilluminated portion of the leaf will be darker, because the chloroplasts are near the upper surface. Desert plants the world over are notably grey, and plants growing in the glaring sand along the sea have the same characteristics in a lesser degree. The fact, that plants do breathe, is given to children without demonstration, and it is not easy to prove conclusively that dust threatens plants with suffocation. Of course they may find out in a general way that plants do not thrive close to dusty roads, and that house plants need to have the dust washed off their leaves frequently.

CHAPTER III

The devices to meet the danger of excessive transpiration are the most striking features of our dry-weather plants. By all means, emphasize the fact of transpiration by letting the children perform the experiments. Ordinary quart fruit jars will do, and if rather large quantities of plants are used, the weights or scales need not be very fine. Of course metric weights are best. If practicable, have the children seal the stems with sealing wax or grafting wax. These simple experiments do not give very definite or accurate results, but the facts that leaves exhale varying amounts of water and that the epidermis restricts evaporation are clearly shown. Of course children can be taught that ordinarily the evaporation of water from plants is invisible, but it is well to make the vapor visible by using the closed jars. On a warm autumn day this requires less than ten minutes. Evaporation soon ceases in the jars because the air becomes saturated, so if the loss of water is to be determined by weight the plants must be exposed to dry air.

As previously stated, water escapes from leaves, not through the epidermal cells, but through the stomata, which are outlets from the intercellular passages. Horizontal leaves have usually their spongy tissue and stomata on the lower side, an arrangement which restricts evaporation since the evaporating tissue is not exposed to so much heat. Vertical leaves, have compact palisade tissue on both faces, and all these leaves with thick cuticle have a reserve supply of water stored in the epidermal cells, and sometimes in several layers beneath them; but these facts can be shown only by means of rather skillful work with the microscope.

The live oak seems to transpire rather freely, but it is not very active during the summer. In Southern California it simply matures its acorns, then appears dormant for several months; in early spring the new leaves come out with a sudden burst. The Eucalyptus seems never to intermit its activity. The extraordinary avidity of its roots for water makes it useful in lands that need drainage, as well as in arid regions. In some species its habit of vertical leaves comes with increasing years, the leaves on young trees being horizontal, and it is very interesting to watch the transition. Many of our native shrubs have leaves of the same type as the oak, so have some common introduced trees, such as the Magnolia and India-rubber; but the pepper tree, and the many acacias have practically vertical foliage. There is a marked tendency to vertical leaves in other native shrubs besides the Manzanita. Perhaps the fact that many cone-bearing trees do well in our climate with little water should be noted. Their leaves, of course, transpire very slowly, and it is for this reason that they can be retained during the winter in cold climates.

The cactus is a type of plant particularly well adapted to regions of long drought for it can remain dormant without losing its parts, above ground. Its roots do not go deep but they are many and long; so during the rainy season the plant can take in water very rapidly, make its periodical growth, and store enough in its aqueous tissue to balance all the evaporation that can occur through its restricted surface during the drought. The plants sometimes look thin and starved at the end of the dry season, but they usually survive. A full study of the cactus is deferred until it is in flower, when the most common species, the tuna cactus, will have true leaves also. Century plants and Yuccas are of the same type as the cactus, and so are many other desert and seaside plants. Introduced plants of the fleshy type usually flourish in our climate. The Russian thistle is an apalling example of this fact in some parts of the state. It should be remarked that the salts dissolved in the cell contents of some of these plants do much in checking transpiration.

The "wild broom," (*Hosackia glabra* or *Lotus glaber*) comes near being a "switch plant." Switch plants, like most Cacti, reduce transpiration by dispensing with leaves, and carry on food-making in the green cells of the stems, but they have numerous slender, woody stems as the name implies. They abound in many desert regions. One plant of this type, *Lepidospartum squamatum*, a strong-scented Composite, is common in the sand washes of Southern California. Its summer leaves are mere scales, but soon after the rains, it puts out ordinary foliage leaves of considerable size.

The Eriogonums, too, are plants that belong to arid regions; they are found only in Western America. The one pictured, *E. elongatum*, is common and typical. In Southern California a still more common one is *E. fasciculatum*, a valuable bee plant, commonly known as wild buckwheat. It begins flowering in the early summer, and will be referred to in Chapter XV. Unlike most Eriogonums, this one has many leaves, but they are hard and small, and, as the name implies, fascicled. They are of a type rather common in arid regions, more common in regions of excessive moisture, and particularly so in regions where these conditions exist in alternation. They are called rolled leaves because the margins of the leaf roll back so as to leave little of the under surface, i. e., the stomata-bearing surface, exposed; the under surface is further protected by felted hairs.

In any part of the state a large number of the plants of this autumn collection will belong to the family Compositæ, a family children

CHAPTER III

can easily learn to recognize because they can usually see that what looks like a flower is really a flower-cluster. Plants of this family are difficult to identify, and the teacher need not worry over ignorance in this direction; but because total ignorance of names gives a feeling of insecurity, the generic names of some common autumn Compositæ are given below. Everlasting plants (*Gnaphaliums*) take their popular name from their dry, silvery flowers; their leaves are usually very strongly scented and fleecy; often the hair is denser on the lower or stomata-bearing surface, but a covering of this sort on the upper surface also is useful in lowering the temperature of the transpiring cells within the leaves. Tar-weed is a name applied to plants of several genera, such as Madia, Hemizonia, and Heterotheca, and in general to Compositæ that are strikingly viscid and ill-scented. One large group of Compositæ is characterized by milky juice and ligulate or strap-shaped florets. The flower represented in Fig. 13, *Malacothriz tenuifolia*, belongs to this group. The plant, in Southern California, often grows six feet or more high, and in autumn it has a very grey, lifeless appearance, except for the beautiful white, pink-lined flowers that are open part of the day. *Stephanomeria virgata* resembles this in general appearance, but is smaller and has heads with fewer flowers inclined to a lavender tint. In many parts of the state the sage brush or wormwood (*Artemisia*)—"old man," the children may call it—abounds. Among the hills one may find golden rod (Solidago), but more likely allied genera (Aplopappus and Bigelovia) with much stiffer habit, asters (either Aster or Corethrogyne), the pretty *Senecio Douglasii*, with large, pale yellow, daisy-like flowers, and fleecy, grey foliage, and sun-flowers (Helianthemum.) Along streams there is a shrubby herb, Baccharis, with willow-like leaves and small white flower heads, and so on,—it would be impossible to give anything like a full list. From this list the introduced weeds, mentioned above, are omitted. The hoarhound—not a Composite—is included in the collection because of its striking adaptation to its environment, for besides its bitter taste and woolliness, it has wrinkled leaves, a device for exposing less leaf surface to the direct rays of the sun.

The turkey-weed illustrates at least five uses of plant hairs;—for defense (not all the stellate hairs are provided with little daggers, only those situated so as to be specially effective), for protection against light, dust and excessive transpiration, and finally for absorbing water and what is dissolved in it. By no means all plant hairs absorb moisture; very often they are filled with air, and water enters them with difficulty. Many hairy leaves, after immersion in water, can be easily shaken dry and have not increased in weight. But it seems to be established that viscid and glandular hairs sometimes attract water to a marked degree. The same viscid substance which in dry air acts like a coat of varnish and prevents evaporation, in moist air may actually attract the water. Kerner states that the glandular hairs of common geraniums absorb water, and believes that, with the water, they take in nitrogenous compounds and even dissolved mineral salts. In the geraniums the water enters the terminal cells of the hairs, the thickened layer of the cuticle here being discarded; in the Eremocarpus the water enters the cells at the basis of the hairs and partially drives out the air from the stellate branches.

The castor oil plant, which is such a flourishing shrub in Southern California, also absorbs water through its foliage, not through hairs but probably through the little warts and cups that have sometimes a very sticky surface. The little warts are under the teeth of the leaves at their very tips; the cups are at the basis of the leaves where ribs converge, and along stems, always on the upper surface. Let the children see for themselves that dew, fog or rain collects at exactly these places. Let them also immerse the blade, not the cut petiole, of a wilted leaf in water and see that it revives.

Plant No. 2, Fig. 15, *Trichostema lanceolatum*, is sometimes called blue-curls because of its long exserted stamens, or camphor weed, on account of the odor, or flea-weed, because of its disagreeable foliage. This plant, too, seems to me to absorb moisture through its glandular hairs. The hairs are pictured at the end of the chapter. The flowers of the Trichostema have a unique method of reserving their honey for bees. The slender tube of the corolla is bent back on itself at such a sharp angle that only a very minute creature could turn the corner. But a large bee, one with a tongue long enough to reach the honey, clinging to the lower petals is heavy enough to straighten the tube and at the same time the essential organs of the flower, anthers and stigmas, are brought with force against the tip of the bee's body. In the younger flowers the stigmas are not mature, but the open anthers leave a goodly supply of pollen on the bee's back. When he visits an older flower, as in the picture, the stigmas strike this pollen and the flower is cross-pollinated. The hive bees are so fond of the honey that they will often be found searching among the fallen corollas in the dust for any honey that may remain in the tube, for hive bees seem to be unable to exhaust the honey from the flowers on the plant.

The other plant in Fig. 15 is *Zauschneria Californica, var. microphylla*, Gray. As it belongs to the fuchsia family, Onagraceæ, its

CHAPTER III

common name is more suitable than these names often are. In Southern California only the variety, the form having small woody leaves, is found in lower altitudes, but in the mountains, where there is more moisture and leaves can afford to transpire more freely, the type with the larger leaves prevail. The flowers are types of those that humming birds love to frequent. They are vivid scarlet, and other guests, excepting long-tongued moths, are excluded because the honey is at the base of a long tube with a marked constriction a little above the base. The birds can usually be seen visiting the flowers, except, perhaps, during the hottest part of the day. As can be seen from the picture, they strike the large, rough stigma first on entering a flower; they are pretty sure to pollinate it if they come from other flowers, for the pollen coheres in considerable masses on account of the cobweb-like threads among the grains and these masses are almost sure to catch on the bird's bill or feathers. If the children watch the flowers they may find that they do not altogether depend upon their guests; the little masses of pollen often swing down so that a light breeze or passing object, sometimes their own fluffy seeds, may land pollen against a stigma. Of course if the children are studying plants for the first time this subject of pollination must be deferred until later. Considering the main topic, the story of the pollination of these two flowers is quite a side issue, but it is a beautiful story and can be read directly from Nature's page.

There are other plants of the dry season that have not such obvious devices for maintaining the balance between transpiration and water supply as the plants we have used for illustration. This is especially true of some introduced weeds. We may attribute their survival to the innate hardiness that has made them universal weeds, or we may see some explanation in their long and often thickened roots, or we may advance the theory that the cell sap contains substances that lessen evaporation, but the fact remains that often, very often, we cannot fully explain these matters. No really scientific person professes to " know all about it," nor is he deeply chagrined when he finds that one of his theories must be abandoned. It is very unfortunate to give children the impression that all of Nature's secrets can be discovered by a superficial glance. But this is no reason why open secrets should not be noted and enjoyed.

The reasons for the fall of the leaves of native California trees in autumn are not easily explained. It can hardly be due to the cooling of the soil, for at just this time the rains cause a general bursting into leaf among small perennials, and even some large shrubs or small trees, like the Ribes and the elder, respond at once to the new moisture in the soil. In the valleys of Southern California the foliage of the sycamores, alders and cottonwoods is practically functionless by October or November, but the leaves do not always fall entirely until after the new ones come in February and March. The alders are in full flower in January. Some willows and the elder are nearly or quite evergreen. The poison oak is leafless for a brief period, but is often in leaf again before January. Introduced deciduous trees take a winter rest, but the leafless period is shorter than in our Eastern States. Such facts as these can be observed, although a theory fitted to them may not be at hand. Of course giving the name periodicity explains nothing. There is really much of interest in connection with the falling of leaves;—the removal of the carbohydrates, and the various alterations in color due to chemical changes that accompany their removal, the fact that the falling of the leaf is a process of excretion, the effect on the soil, etc. Leaves are not broken off leaving a raw surface subject to injury, but a layer of separation, consisting of corky cells that separate easily, is formed, and the leaf scar is waterproof from the first. The buds in the axils of the leaves, or under the petiole in the sycamore, should be noted. In short, everything possible should be done to foster or to cultivate in California children a love for trees.

CHAPTER IV.

SOME PLANTS THAT DO NOT MAKE THEIR OWN LIVING.

This chapter deals with parasitic and saprophytic plants of all ranks. Of course any real study of bacteria is not to be thought of in elementary work ; but although the subject must be treated purely in an information way, its natural interest and practical bearings seem to justify giving it some attention. Bacteria are very minute unicellular plants. It would take about half a million of some kinds to cover the dot of an " i " on this page. The plants reproduce by cell division with almost incredible rapidity. It has been estimated that, under favorable circumstances, one plant can give rise to sixteen millions in eight hours. This fact explains the rapid progress of some germ diseases, the cholera for instance. Some bacteria form spores, and many of them offer considerable resistance to adverse conditions. Some can survive even freezing or boiling temperatures and very long periods of dessication. The universal presence of bacteria is much less alarming because of the fact that many species are harmless.

Bacteria that have not access to free oxygen in the air, have the power of obtaining it by breaking up fluid compounds that contain oxygen. In this way they may rob the blood of oxygen and bring about other decompositions that are supposed to be the causes of some diseases. Nature seems to have two methods for counteracting the effects of these injurious bacteria. Animal bodies contain cells that can digest bacteria ; they also contain, it is said, certain substances termed defensive proteids that can destroy bacteria or their products. So immunity from certain diseases can be secured by one of two artificial methods. The bacteria that cause the disease can be cultivated until there results what is called an attenuated virus ; inoculation with this virus leads to a very mild form of the disease and prevents a virulent attack. The other method is to introduce the defensive proteids into the system.

SUPPLEMENT

Bacteria that cause diseases enter the system by different avenues, through abrasions of the skin perhaps, or by way of the lungs or the alimentary canal. Because our state is to such an extent a sanitarium for consumptives, it is well to emphasize the fact that the chief danger of infection is from bacilli liberated in expectoration, a danger that can be met by proper care and cleanliness. Typhoid fever arises usually only when the germs enter the alimentary canal, and is therefore a preventable disease.

The bacteria that cause putrefaction split up albuminoid compounds into various gases and liquids. In the earlier stages some of the products have very offensive odors; later on nitric acid, so essential to plant life, is evolved. The unused parts of carbo-hydrates also are returned in various forms to the surrounding air or water. There are other very useful bacteria whose presence in the roots of leguminous plants is denoted by warts or tubercules. They are apparent even in small seedlings, in bur-clover or lupine for instance. It seems to be well established that in this case the bacteria can fix and store up free nitrogen for the use of the higher plants; that is, this is not a case of parasite and host, but is true symbiosis. The souring of milk and the ripening of cream are also due to bacteria. There are, in common daily experience, many other evidences of the presence of bacteria. The cloudiness of water in which cut plants have been standing is a visible evidence, so are the bright red spots often seen on fermenting starchy foods.

Yeast plants are nearly as minute as bacteria, and their study requires high powers of the microscope. They split up grape sugar into alcohol and carbon dioxide, thus causing alcoholic fermentation. They can reproduce very rapidly by a process called budding, sending off little sprouts or buds that break off and become independent plants. The facts that yeast plants have another method of reproduction, and that their affinities are doubtful, need not be discussed here.

Mould can be grown with very little trouble. Bread moistened and closely covered will grow a good crop in four or five days. Usually the growth is so vigorous, that not only the substratum of food, but the sides of the dish and the cover as well, become covered with tiny clusters of spore-bearing stalks, connected by runners. These clusters are clearly perceptible without even a lens, for the stalks are several millimeters long. They are easily handled, and a low power of the microscope should be used. It is best to mount in glycerine, adding a drop of alcohol to expel air. The filaments of the mould are not divided into cells. From the first they branch or interlace, forming

CHAPTER IV

a cottony mass. Soon, at definite points, root-like branches are sent into the substratum, and the clusters of upright stalks appear. Each one of these stalks enlarges at the upper end into a little sphere into which dense protoplasm collects; finally a convex partition converts this swelling into a cell, and the protoplasm divides into many spores. Ultimately the outer wall of the cell dissolves and the spores float off in the air, leaving only the stalks and the collapsed inner wall or partition. This is, of course, simply cell division, but there is sometimes a union of two cells to form a resting spore. So in reproduction as well as in structure the mould resembles some filamentous algæ. Some botanists regard fungi of this type as degenerate algæ that have lost chlorophyll and the power of making their own living. But the relationship of fungi to other plants, and their affinities among themselves, are matters that are far from settled. Not only this, but the life histories of many fungi are not known. The whole subject is a difficult one, and genuine work with it is undertaken only by specialists. Every civilized government recognizes the value of work of this kind in the department of agriculture, and has in its employ men whose lives are devoted to this branch of applied science.

It is easy to impress children with the importance of this subject and to stimulate their interest in it, by frequently calling their attention to these fungi when they are visibly present on common plants. Smut, for instance, on the flowers, (that is, on the tassels and ears) of corn is of common occurrence. The life history of smut has been only recently known. The black sooty mass consists of spores, at least they are commonly called spores, but they do not germinate directly on corn plants. They fall to the ground and send out branches that produce quantities of very minute spores. Even these spores cannot penetrate a mature corn plant; they simply infest the ground and lie in wait for seedlings. They can enter the tender tissues of the seedlings, and, germinating there, they produce a characteristic fungus mycelium, as this mass of vegative cells is called. In this case the filaments of the mycelium are so exceedingly delicate that it has been difficult to detect their presence in the most careful microscopic work; but it has finally been discovered that as the corn plant grows, this invisible foe steadily advances to the parts whence the flowers are to arise. When the plant sends great supplies of nourishment for the development of its own essential organs, the fungus suddenly springs into great activity and uses these supplies for the production of its own spores.

Rose leaves in our gardens are likely to be infected with rust whose spores are found on the under sides of the leaves. At first only the

SUPPLEMENT

red spores appear, later brown or black spores show among the others, often quite hiding them. **Under** the microscope **the red spores are seen to be** single cells with **thin walls, but the dark spores consist of several thick** walled cells **raised on a stalk.** This illustrates a fact that botanists were long **in finding out, the** fact that many rusts produce different sorts of spores **at different** stages. It is also true that the same fungus may, **during part of its existence, live on one sort of plant, and later on require quite a different plant for its host. It is certain, for instance, that one kind of** rust must spend its first period **of life on barberry bushes,** and that it produces there spores in cluster cups, that is in little cups imbedded in the **leaves.** These spores will **germinate only when they fall on wheat.** The mycelium within the tissue of the wheat produces great quantities of red spores, which, in turn, germinate immediately and produce **more red rust, until the whole grain field is infected.** As the grain ripens, the same mycelium that produced thin-walled red spores **now** produces the thick-walled, **darker** spores that can remain dormant during an unfavorable season. At the beginning of the growing season these spores on the rotting straw germinate, and, like the spores produced by smut on corn, produce great numbers of minute spores; these spores germinate **only when they fall on the barberry, and forming cluster cups, complete the cycle. It is an interesting fact that, long before the life history of** wheat **rust was known, it was believed that the barberry exercised an evil influence over wheat, and in 1670 the** Massachusetts **legislature passed laws for the extermination** of the barberry. Probably not all **rusts pass through as many stages as the wheat rust.** In fact it is **pretty certain** that **the rust so common on** the **malva** has neither the cluster-cup nor the red-rust stage. Examples of the cluster-cup stage **are more likely to be found** in the spring months. In Southern California **they are** common on *Phacelia distans* **and on nettles.**

The life histories of fungi that affect agricultural products, provided they are known, can often be obtained by applying to government headquarters. Frequently the stories of discovery in the fields of research are of absorbing interest. The life **of Louis Pasteur by his son-in-law is** a striking example.

It should be possible anywhere in California to obtain sufficient material to arouse an interest in lichens. A supply of the larger, more striking kinds from the coast or mountains, once obtained, will last for years; but the smaller, more common kinds will answer, and children should soon be able to find lichens for themselves on fire wood, rocks, walls or trees in shaded places. Special effort should be made to find specimens with spore-fruits; these are usually cup or

CHAPTER IV

saucer-shaped, and have a smooth, almost polished inner surface, often of contrasting color. The nature of the fungus part is made more or less apparent by loosening the lichen from its substratum, or by picking it apart. If lichens in any considerable quantity can be obtained, their avidity for water can be shown by measure or weight. Some lichens absorb fifty per cent of their weight of moisture in ten minutes. Water often dissolves the pigment of the lichens, and usually it renders the green color of the layer of algæ just beneath the upper surface more apparent.

The remarkable symbiosis between the two kinds of plants, the fungus and the algæ, which make up the lichen, has been discovered only in the latter half of our own century, but it seems now to be generally accepted as a fact. Full demonstration of it could not, of course, be undertaken in an ordinary school-room, but the interweaving hyphæ of the fungus and the imbedded algæ (or more properly green plants of even lower rank) can be shown under a moderately high power of the microscope. More skillful manipulation is necessary to show the spore-sacs clearly. Any one attempting this for the first time would better begin with the spore-fruits of the "cup fungus," or Peziza, which is nearly related to the lichen fungus, and is neither tough nor brittle. The spore-fruits of the Peziza are rather common in manure during the rainy season; the most common kind is a light brown or amber cup, from one and a half to two inches in diameter; the inner surface of the cup is velvety, and a vertical section shows that every thread of the pile of the velvet is a slender, delicate sac, many sacs containing eight spores. If you keep these spores under observation, you will see that on maturity they suddenly expel little clouds of spores. The sacs that form the lining of the lichen cups are much smaller and have their tips hardened, that is, it is their hardened tips that form the polished inner surface of the cup. They, too, expel their spores at maturity, and bodies so minute and in such multitudes are sure to be scattered everywhere by the winds. These spores, of course, belong to the fungus part of the lichen, but myriads of the one-celled host plants in their resting stage are also blown about. A piece of moist blotting paper exposed to the wind, even on a high mountain top, will collect a goodly number of both fungus spores and host plants. The coarser, meal-like powder often seen on lichens, consists of fragments, little colonies of algæ and fungus combined, that are capable of growing into independent lichen bodies.

It is hoped that the subject of nutrition is made clear in the Reader. The green host plants act as green cells always do in making organic food. The delicate cells of the fungus mycelium absorb food very rapidly, the food material being mainly atmospheric moisture and the salts and gases dissolved therein, but the hyphæ that penetrate the substratum do select and appropriate from it some food materials, which they dissolve by means of an excretion. Lichens, like many other lower plants, and like liverworts and mosses in a higher group, have the power of surviving long periods of dessication and reviving with marvelous rapidity.

Children are sure to be interested in toadstools, and there is no difficulty in getting a collection of them after a heavy rain. Urge the children to hunt for the part that absorbs the food, the mycelium. It is sometimes almost invisible, the delicate filaments or hyphæ being obscured by the soil. When the mycelium must thread its way through loose, decaying leaves, it is sometimes a webby mass of considerable extent, but usually the vegetative part is absurdly small as compared with the spore-bearing part. The rapidity of the growth of the sporefruit is simply inexplicable. We may bear in mind that rich, readymade food is at hand, and that the long, slender, thin-walled cells of a fungus mycelium always have the power of rapid absorption, or that the mycelium may have been at work for a long time before sending up the spore-fruit, still the cell-making energy that results in the formation of a compact body as large as an egg from a minute, vegetative part in a few hours, is marvelous.

The collection of toadstools is pretty sure to include the one most commonly eaten, *Agaricus campestris*, and this species shows very prettily the different stages of growth of the cap or spore-bearing part. The little white nodules early show the beginnings of gills within ; then there is the stage where the veil, or volva, which has protected the gills, is beginning to break from the stalk. Have the children notice that in this species, the remains of the volva always form a ring well up on the stem. In the poisonous Amanita, No. 2, the broken volva is left at the base of the stalk forming a little cup. As the cap of the Agaricus expands, it can be seen that the gills are unequal in length ; they are at first a delicate flesh color, but they become darker as the spores mature, the ripe spores being nearly black.

The collection will also probably contain some of the Coprinus group, toadstools whose caps do not fully expand and that liquify when the spores are mature. The large one, No. 3, in the picture, is a delicate grey and pink species ; small light-brown Coprini are more common. Sometimes members of this group have a rather fœtid odor that is useful in attracting carrion beetles, and flies.

CHAPTER IV

The number of spores produced by one toadstool plant is beyond computation. The stalks that give the velvety appearance to the gills are so densely crowded and so minute that very thin sections and high powers of the microscope are necessary to show them clearly, and each stalk bears four spores. These spores may, of course, be scattered by the wind in addition to the other methods of dissemination suggested. Mushroom "spawn" from which the cultivated *Agaricus campestris* is propagated is the mycelium of the fungus.

The value of edible fungi as food can hardly be overstated. They have nearly the same value as meat, a fact not surprising when one reflects that fungi in their methods of nutrition resemble animals rather than green plants. In older countries, like France and Germany, where the struggle for existence is keener, this great waste of food does not occur. The peasants gather and preserve great quantities of toadstools. In our Eastern States the edible kinds have been described and illustrated, both in popular works like Gibson's charming book, and in government publications. On our coast it is not so easy to learn to know the edible species. Prof. A. J. McClatchie has tested the species about Pasadena, with the conclusion mentioned in the Reader, but the results of his work have not been put in form generally available. When the intelligence of the public demands it, probably ways and means will be found to make the identification of our edible fungi, too, a matter of certainty.

It cannot be too emphatically stated that there is no infallible rule for distinguishing all poisonous from all edible species. We must learn to know each edible species so as to surely distinguish it from all others, just as we know certain species of higher plants. It is true that the edible kinds greatly outnumber the poisonous ones, but the poisonous species of the Amanita group are usually fatal. There are numerous species not poisonous that are not edible on account of an acrid or otherwise disagreeable flavor.

As stated in the Reader, some puff balls are edible, and so are some species of other fungi nearly related to toadstools and puff balls; but we shall have to wait for further knowledge of our California fungi before we can fully avail ourselves of our resources of this sort. In the meantime there is much that is attractive or interesting about these allied forms. The spores of some of our puff balls are intermixed with minute filaments that are hygrometric and by their movements aid in distribution. The earth-star of the picture is common in oak-covered slopes in Southern California. As stated, the outer coat closes up in dry air; I am told that some other earth-stars close in moist air instead. The shelf-fungi are sometimes very pleasing in form and coloring. There are other sorts of Polyporus that are extremely mischievous, forming the so-called dry rot of timber in dark cellars, mines, etc. The "stink-horn" fungus is rather common in the vicinity of Los Angeles. When it first appears above ground, it is enclosed in a brown or purplish membrane, and looks rather like a potato ; but soon this membrane bursts, and the spore-fruit, in the form of horns with a honeycombed surface, appears. The fœtid odor is overpowering; it serves to attract flies, which feed on the sticky fluid surrounding the spores and thus carry many of them away.

CHAPTER V.

AFTER THE RAINS.

The time for this lesson in the average year in Southern California is December. It should not be taken up until the seedlings and early perennials are well started, and are in the characteristic condition of plants that have a good supply of water. As in the lesson on autumn plants, this chapter in the Reader should, by all means, be preceded by a field-day lesson conducted by the teacher. Perhaps after these two excursions by the class as a whole, children can be trusted to collect material for other lessons by themselves. If this field lesson can be undertaken before the holidays, it will be a stimulus to the children to undertake like expeditions by themselves during the holiday vacation. It would be well to call for a collection of ferns, liverworts and mosses for the first lesson after vacation. The liverworts, mosses, and ferns growing from prothallia (see Fig. 30,) should be shown the class on this excursion. The experiment with the bur-clover seeds, too, should precede the use of the Reader. Put some free seeds, also some burs containing seeds, on moist sand, cover with glass and keep warm. The burs should be somewhat anchored by sprinkling a little sand over them, so as to imitate the condition of the wayside burs. The experiment will require three or four days, and it is well to first soak the seeds and burs a few hours.

This chapter of the Reader needs little exposition. The first part is but a summary of what the children should have already observed. The question of the use of the clover seeds remaining in the bur, should have been proposed, but not answered, except as the children have thought out the answer for themselves. A main physiological point of the lesson is to show the behavior of plants that have plenty of moisture but not much heat. Obviously a stretching or spreading out to the sun will be a most striking phenomenon. **Malva** leaves face the sun persistently, and, on account of their long petioles, easily avoid shading one another. The bur-clover, *Medicago denticulata*, and the filaree, *Erodium cicutarium*, and *E. moschatum* are equally successful. *E. cicutarium* is more likely to form pretty leaf rosettes, but *E. moschatum*, in the vicinity of Los Angeles at least, grows more luxuriantly; the musky, ferny, leaves grow close together, every one erect and rejoicing in the sun, and they form a most exquisite covering for our waste ground and waysides.

The subject of the arrangement of leaves on the stem, or phyllotaxy, has been worked out very fully. The spiral arrangement that often exists can be expressed, it has been shown, by a series of fractions that progress in the most fascinating way; but this arrangement can be detected only in vertical shoots that grow under simple conditions. Ordinarily on the same plant vertical and horizontal shoots will present different arrangements on account of the twisting of the leaves to accommodate themselves to the light. If a vertical shoot be fastened down to a horizontal position, the leaves will adjust themselves to the change. Kerner in his "Natural History of Plants," pages 392-424, and Lubbock in "Flowers, Fruits and Leaves," have discussed this subject in a most entertaining way. In Newell's Reader in Botany, Part I., extracts from both are given. But the details of the subject are much too complex for children; besides, these observations, made mainly on the trees of England and of Central Germany, assume that the arrangement of leaves is to secure the greatest amount of illumination possible, and this is true of California plants for only a few months in the year. Our trees and perennial summer plants present a notably different appearance from the the trees of countries with a short, moist, warm season.

But at this season our new leaves have many striking devices for getting as much sunlight as possible. Illustrations are everywhere. It is a good plan to have the children try to sketch leaf rosettes and mosaics, looking down on the plant of course. The leaves of nettle seedlings are arranged on a simple plan and lend themselves well to this exercise. So do climbing plants, for they are obviously most likely of all to form mosaics.

In the valleys of California, native vegetation seems to be in little danger from cold, and winter plants with hairy or woolly coverings to protect them against sudden changes are not very common. Nor, so far as I have observed, is there a large number of plants that take "sleeping positions" at night. Beside the clovers, the oxalis, wild and cultivated, and the common introduced acacia with bipinnate leaves, are well known examples. The oxalis will take this position when placed in a dark box, and the acacia on being picked or roughly handled, so the folding of these leaves can be observed or drawn at

CHAPTER V

any time. There is no doubt of the value of this habit to the plant in preventing radiation of heat from the leaves. Darwin found that clover leaves artificially prevented from folding on a cold night perished, while the other clover leaves survived. A discussion of the causes that produce these movements of the leaves, hardly belongs to elementary work.

It is not strange that a climate like ours should have a large number of perennials with bulbs or other underground storehouses. These underground parts can store more food than seeds, and are protected by the soil from evaporation during the long dry season. They need only to protect themselves against the attacks of rodents as most of them do. I have found that the soap-root, *Chlorogalum pomeridianum*, appeals especially to little folks, perhaps because of its utilitarian associations. Its botanical story is easy to read. The flowers open at night and perhaps stand a better chance for pollination by coming so late in the season. The cluster lily, *Brodiæa capitata*, and the blue-eyed grass, *Sisyrinchium bellum*, are such common plants that they should be known from their first appearance. Common names for the Brodiæa are many; calling it wild onion should be discouraged because of the confusion with the true wild onions, *Alliums*, that are common in many regions. Of course Sisyrinchium is not a grass, but the common name is well established. Many of the perennials of this season belong, like No. 1, to the family Umbelliferæ; the mint family, Labiatæ, too, is well represented in the new growth of this season. Of course the ferns are not to be ignored in this field lesson because they are to be studied later. There is no reason, except the matter of time, why they should not all be taken up at once. Children should certainly know the common names of all the ferns of their vicinity.

The Chilicothe, Micrampelis, or Megarrhiza in California Flora, deserves, I think, all the space given it in the Reader. It is a common and an attractive plant and is interesting in many ways. In Rattan's "California Flora," an account of the germination of its seeds is given; children should be asked to collect the seeds during the summer for growth in the schoolroom another year, as the seeds seldom germinate out of doors in Southern California. Perhaps some one else will succeed better than I have in seeing insects pollinate the flowers. I have seen only flies visit them in a haphazard way that did not seem to benefit the flowers. I have not had opportunity to watch the plants at night. Of course the showy, white corollas are superfluous if the flowers are pollinated by the wind only. If a plant should be growing near the school building, by all means

SUPPLEMENT

have the children watch the climbing **carefully. The** subject **of** climbing plants is one of great interest, and will be referred to in other **chapters**; the attention of the children **should be directed** to the subject at every opportunity. In this case, **if practicable, mark a** young tendril, note how it grasps an object, **when it coils, etc.**

Perennial lupines, **as** well **as the annual** species, are quick **to** respond to the rains. In the vicinity **of Los** Angeles, *L. albifrons* blooms **all the** year on northern slopes, but the flowers are much more abundant during the rainy season. The wild currants (*Ribes glutinosum*, **or** kindred species) are in full flower in December in the foot-hills of Southern California, **and are common ea**rly flowers throughout the **state.** Their fancied resemblance **to the** trailing arbutus is of course only in color and form. The flower clusters are more or less pendent **to escape** wetting—moisture is fatal to pollen—and often the cluster is exactly beneath a leaf, but the protection of the flowers by leaves is **not so** marked as in **the** gooseberries. I have seen the flowers pollinated by bees, hive bees and the larger native bees, also by butterflies. The stigmas I have noted have been usually, but not always, slightly beyond the anthers. They are held rigidly in the entrance to the flower and are sure to be struck by the tongue of the entering guest. The anthers, which **open inward, are ranged round the narrow** entrance, **and are likely to be struck also. I** have seen undoubted cases **of self-pollination when the anthers and stigmas were at equal** heights, although when the flower first opens the anthers are held at a slight distance from the stigmas. I find the plants thronged with **bees when the weather is** at all favorable.

The gooseberries that flower in December and January in the San **Gabriel mountains, are** *Ribes amarum* and *R. hesperium*, species allied to *R. Menziesii*, which is common in other parts of the state. **They are beautiful,** graceful flowers, with their long, red calyxes, reflexed sepals, and white petals, and they seem to me to be mainly pollinated by humming birds. *Ribes aureum* or *R. tenuiflorum*, the yellow "flowering" currant, and *Ribes speciosum*, a very common later scarlet gooseberry, will be referred to again. The wild **lilac** Ceanothus, **will also be** considered later, when the children are better prepared **to understand its pollination.** *Ceanothus crassifolius* flowers early, and has rolled leaves with furry backs **that assist tran**spiration now **by** keeping the dew from wetting the stomata, and later on, check transpiration, as has been shown before.

You are likely to find several deciduous shrubs that, like the Ribes, **require** little heat to bring **out** the new **leaves, and these leaves are almost** sure to present some **features of special interest.** Several

CHAPTER V

have a bloom on their under side, which sheds the dew at the slightest disturbance. Do not fail to have children notice that dew collects on the lower, as well as on the upper side of leaves. The fact that dew remaining a long time on plants, hinders transpiration, will be brought out in the next chapter. *Pentstemon cordifolius*, a plant sometimes called wild honeysuckle, pictured in Fig. 71, puts out leaves in early winter. This is a climbing, or perhaps more properly, a weaving plant, and it is easier to see its method of getting up in the world now, than when it is in full foliage. It sends out some new shoots that seem to be pushing upward for support, but shoots from branches that have already looped themselves over a supporting twig of some other shrub, often continue on in a horizontal position, and the different disposal of leaves on these two sorts of shoots, is very marked indeed. The leaves are opposite, and on the vertical stems they stand at right angles with the stem, in four orderly lines, an arrangement advantageous for short petioled, broad-based leaves, so long as stems are vertical. But when the shoots are horizontal all of the leaves twist on their petioles sufficiently to face upward, and the leaves of the alternate pairs bring themselves into the same plane as the other pairs, neatly filling in the interspaces. This is a common habit of plants with opposite leaves, and it will not be difficult to find other illustrations. The fact that the poison oak sometimes climbs like the English ivy, by aerial rootlets, should be noted. The new leaves and shoots of the poison oak, Pentstemon, and many other plants, have a very noticeable amount of red coloring matter, or anthocyanin. As stated before, this color in stems and veins is supposed by some botanists to act as a screen, protecting the various products that pass along these routes; and it seems to be fully proved that, by changing light to heat, anthocyanin is of special service to young leaves.

By this time the buds on many of the trees may be somewhat swollen, particularly on the willow, sycamore and alder, and the alder is probably in flower. It is easy to see why it is an advantage for the flowers of wind pollinated trees to develop before the leaves. The willows are pollinated by bees as will be shown later. It is very desirable indeed for children to take a first lesson in pollination from the willows. Much of the observation called for in the chapter on the "Awakening of Trees," must be carried on beforehand. The teacher should, if possible, select trees that can be watched without special excursions; the next best plan is to have twigs that show signs of activity brought to the schoolroom and kept in water.

SUPPLEMENT

CHAPTER VI.

FERNS AND THEIR RELATIVES.

The plants of this chapter constitute the group Archegoniatæ, a name suggested by their method of reproduction; but while their peculiarities of reproduction are of special interest to the biologist, and form properly a basis for classification, the subject is not within the comprehension of children, and so is hardly touched upon in the Reader. A thorough study of the reproduction of ferns and all higher plants demands skillful work with the microscope, but perhaps all teachers should have some general idea of the subject.

This division of the plant world, Archegoniatæ, is the highest among flowerless plants, or, more properly, among plants that do not produce seeds. It includes two groups, the Bryophytes, comprising mosses and liverworts, and Pteridophytes, that is, ferns, horsetails, club-mosses, and a few others. All members of these groups bear organs called antheridia, which contain fertilizing cells called sperm cells, or spermatazoids; also organs called archegonia, which contain egg cells or oöspheres. As in seedless plants generally, the fertilizing cell swims to the egg cell; it is only in seed-bearing plants that the fertilizing cell, the pollen grain, is transmitted through the air. Necessarily, then, all Archegoniatæ must inhabit moist places. After the union of sperm and egg cells, there grows a more or less complex body that produces spores. From these spores grow the plant bodies that bear archegonia and antheridia, and so on; so there are always two phases—two generations in the old terminology—in the life of these plants : from the spores grow the sexual plants ; from the union of two cells produced by these plants grows the spore-bearing plant, or sporophyte, as it is sometimes called.

The plants that grow immediately from liverwort spores are usually similar to Nos. 9 and 10, Fig. 30, that is, they are horizontal leaf-like bodies fastened to the soil by root-hairs. The antheridia and archegonia are imbedded in these bodies; sometimes both kinds of organs

CHAPTER VI

in the same, sometimes in different plants. After the sperm cell swims down the archegonium and unites with the egg cell, there grows from the fertilized cell, a spore-bearing body; in the liverwort, No. 9 in the illustration, this body is like a little umbrella; it is sometimes urn-shaped, sometimes a mere sphere beneath the frond. This spore-bearing body remains attached to the mother plant, and is mainly supported by it, though it may assist in food-making. The moss spore produces first a branching, filamentous body resembling some of the green algæ, but soon, from buds on these filaments numbers of tiny plants arise, each consisting of stem, root-hairs and leaves, that is, the part of the plant that is commonly noticed. The antheridia and archegonia are produced among these leaves, sometimes in the midst of leaf rosettes which resemble flowers. After an egg cell in an archegonium has been fertilized, the spore-bearing body begins to grow; it is usually urn-shaped and borne on a stalk. As in the liverworts, this spore-bearing part remains attached to the sexual plant, but it has always green cells and stomata and helps in food-making.

From the fern spore grows the prothallium, which, in all our species, much resembles the first generation of the liverwort. The prothallium bears archegonia and antheridia, and from an egg cell, after the sperm cell is united with it, grows the plant we commonly call the fern. This spore-bearing plant becomes capable of nourishing itself very early in its existence, and the sexual plant perishes. So the first generation of the fern is comparatively minute, and has a brief existence; the second generation is of considerable size, it is as well equipped as flowering plants for supporting itself, and it may live for years. The first generation of the horsetail, Equisetum, is even smaller than the fern prothallium, while, as we have seen, the sporophyte is a complex plant of long duration.

Perhaps the most interesting Pteridophyte, biologically, is a small club-moss, Selaginella. One species is common on California foot-hills and mountains. It might be mistaken for a very hardy moss; its stems are woody and much branched and its closely crowded leaves are as small as moss leaves but much thicker. The Selaginella produces, in the axils of leaves, two kinds of spores, macrospores and microspores, that is, large spores and small spores. Four of the large spores fill a macrospore case, while a microspore case contains many spores. Both kinds of spores fall to the ground and germinate. The prothallium growing from the macrospore always produces archegonia containing egg cells, but the prothallium is minute and merely protrudes from the thick spore wall, which also encloses a store of

SUPPLEMENT

food for growth of the young sporophyte; so it is apparent that this macrospore is analagous to an ovule, which also contains, besides stored food, an egg cell surrounded by a tissue more or less similar to archegonium and prothallium. The prothallium produced by the microspore consists of but one cell, and bears but one antheridium containing spermatazoids; so the microspore approaches the pollen grain in simplicity of structure, for there is one portion of the pollen grain that is supposed to correspond to a spermatazoid.

The ovules of the Coniferæ are more nearly analagous to macrospores than are the ovules of higher spermaphytes, but a thorough study of the fertilization of seed-bearing plants requires more skill than a like study of lower plants; so the old term Cryptogram, meaning hidden fertilization, is really more applicable to higher than to lower plants. On the other hand, the transference of the fertilizing cell, that is pollination, is easily observed in seed-bearing plants, while the corresponding process in lower plants always occurs in water and can be seen only with the aid of the microscope.

The lowest of Archegoniatæ, the liverworts, vary in form. Some of them are as thin and minute as fern prothallia, which they closely resemble, while the higher ones, the scale-mosses, resemble mosses, having slender stems and thin, green leaves; those in the illustration are of the type most common in California. No. 10, Fig. 30, the Lunularia, is a common green-house pest; like many other plants growing under very favorable conditions, this liverwort reproduces only by fragments; the fragments in this case are tiny green buds, called gemmæ; they are contained in the crescents at the tip of the frond. No. 9 is the female plant of a very common, native liverwort *Fimbriaria Californica*. This liverwort is perennial; like many other California liverworts, it survives the dry season by giving up its moisture and curling its upper surface inward until it seems a fossil, rather than a dormant plant. With the first rains the plants expand at once, and colonies of them, usually forming rosettes, are common and rather conspicuous in springy, or moist, shady places. The plants of the female colonies have at first a nearly smooth upper surface, then a tiny ball appears, and they ultimately become like the one in the picture; the male plants have ridges on the upper surface. The spores of *Fimbriaria Californica* are very interesting, and are easily shown with low powers of the microscope. They do not mature until the summer months, but in winter some can usually be found still adhering to the old capsules. Shake out the coarse, yellow powder on a dry slide, moisten very slightly by breathing on it, mount without cover glass, and examine quickly. Besides the large, rough spores there are long cells, called elaters, with spiral markings. As the elaters dry, they coil up suddenly with considerable force and so scatter the spores.

Mosses do not generally form a conspicuous part of California vegetation, but smaller kinds can always be found in moist weather. The moss frequently chosen as a type in text books of botany, *Funaria hygrometrica*, is very common in California. It will often be found even along city streets, on gravelly soils shaded by hedges or walls. Its "flowers" are abundant in winter, that is, its clusters of yellow or brown antheridia are plainly visible in the centers of many of the plants, while other plants have archegonia and spore-capsules in various stages. The antheridia are easily mounted, and a high power will usually show escaping spermatazoids. Archegonia are not so easily found. Ripened moss capsules, or spore-cases, are exceedingly dainty and interesting objects. After the calyptra, which is the remains of the archegonium, and the operculum, or lid, have been thrown off, there are still teeth that act hygrometrically, that is, open and close with varying degrees of moisture, and so control the dispersal of the spores. Children should, of course, be encouraged to collect the larger and prettier mosses when they are available.

In spite of the long droughts of California, wherever there are shaded nooks we are pretty sure to find some of our hardy ferns. Ferns can be made a very impressive text for teaching the relation of climate to vegetation. Like liverworts and mosses, they must have moisture for fertilization, but they cannot endure entire dessication as can the other two groups. The nature of their foliage, also, and other conditions, render a certain amount of moisture imperative; so in our country, and in others, there are areas of level, treeless lands, hundreds of miles in extent, quite destitute of ferns. On the other hand ferns are specially adapted to regions of excessive moisture. We have some narrow, sunless cañons that illustrate this, but generally children will have to get this idea from greenhouses and from pictures. Kerner, speaking of the mountain regions of Jamaica, says: "Here are found some five hundred ferns and a large number of mosses and liverworts. The level or sloping ground, rocks, the forest floor and decaying tree trunks, all are covered with ferns of every shape and size; there are groves of tree ferns, the trunks of trees are invested, right up to the crown, with delicate green fronds, whilst tiny representatives of the filmy ferns have actually taken up their abodes on the foliage leaves themselves."

Of the ferns mentioned in the Reader, *Polypodium Californium*, Kaulf., is very common in the coast ranges. It grows rapidly and matures spores during the winter season. It belongs to a large family that adapts itself to many climates and conditions. Kerner tells of polypody leaves in a dark cañon that grow five feet long. The goldenback fern, *Gymnogramme triangularis*, Kaulf., is widely distributed in California, but the silver-back, which is merely the variety *viscosa*, is limited to the southern counties, being apparently better adapted to dry habitats; it is common on Catalina Island and in the mountains of San Bernardino and San Diego counties. The most common maidenhair fern in the coast ranges is *Adiantum emarginatum*, Hook.; the delicate Venus-hair, *A. capillus-veneris*, is found along waterfalls and springy places in Southern California, and the fan-shaped frond of *A. pedatum* is found occasionally in the south and rather commonly farther north; the last species is common in our Eastern States, and in cultivation here. The coffee fern, *Pellæa andromedæfolia*, Fee., and the bird-foot fern, *P. ornithopus*, Hook., are widely distributed and hardy. They often survive the summer without dying down, but their main growth is during the winter season. The lace and woolly-back ferns, various species of Cheilanthes and Notholœna, inhabit either dry hillsides or high altitudes. The bracken, *Pteris aquilina*, Linn., is the most widely distributed of all ferns; it seems to be, even in our climate, a summer rather than a winter plant; so in Southern California we find it in the cañons, along streams, where it attains a considerable height, or in pine woods in the mountains, where it is less luxuriant. *Woodwardia radicans*, Smith., which is mirrored in the pools of so many of our mountain streams, is a veritable tropical fern, being also native to Peru, Abyssinia, India and Java. Every California child should know this fern of ours, and in most localities it is probably possible for the teacher to obtain cut leaves at least. This fern may be successfully grown as a garden plant. One of the shield ferns, *Aspidium rigidum*, Swartz., or the variety *argutum*, Eaton, is common in well shaded places, and is sometimes so luxuriant that it is mistaken for Woodwardia. Another shield fern, with simply pinnate leaves, *Aspidium munitum*, Kaulf., No. 5, Fig. 28, is common in the north and also in the higher altitudes in the south; it is graceful and very hardy, and is frequently seen in cultivation. The lady fern, *Asplenium felix-fœmina*, Bernh., is always a favorite; it is not rare in northern woods, but it is seldom found in Southern California.

There are excellent reasons for encouraging children to become acquainted with all available native ferns; ferns are naturally attractive, and are very easily and satisfactorily preserved; their habits are marked and interesting, and their adaptations easily read. If the order of the Reader has been followed, an entire fern plant, such as

CHAPTER VI

the Polypodium, with its numerous fibrous roots, strong root-stock, and prominent woody bundles and veins, should be made a text for reviewing the physiology of higher plants, especially the functions of woody tissue. The habits and adaptations of the ferns are perhaps sufficiently emphasized in the Reader. Fern leaves are very cleverly disposed in the bud, the delicate tissue is infolded, and the stronger, woody tissue forms a spherical screen about it ; at the same time the petiole, or its continuation, the mid-rib, is an arch that pries up the soil, then straightens itself and draws out the unfolding leaf. The devices for guarding against both too much and too little evaporation are similar to those possessed by other plants in the same habitats. The reason for the immunity of ferns from the attacks of animals is not so obvious. They are also said to be little troubled by fungi, but the golden-back is sometimes infested with a rust.

The spore-cases of ferns are remarkably attractive objects under the microscope ; they are very easily mounted and do not require high powers. The Polypodium in February is likely to have spore-cases in all stages of development, those forming the darkest brown spots, being, perhaps, cases that have already thrown their spores. It is not really at all strange that the growth of ferns from the spore should have been unknown until the days of microscopes. The superstitions about fern "seeds" are quite characteristic of the science of the Middle Ages. There was, indeed, one skeptical spirit who proved that one need not go alone nor use magic spells to collect the "seeds," but he did not dream of proving by experiment whether or not they were seeds. The arrangement and coverings of the spore-cases, or sporangia, form a basis for the classification of our ferns. As has been noted, some sori, as the collection of sporangia are called, are naked, others are protected by rolled back margins, others by a special membrane called the indusium.

There is an exquisite little floating plant, the Azolla, sometimes called the water fern, that is common in still water. Single plants are not more than half an inch long, but in the mass their tints of pale green, purple or almost red, are very attractive. The reproduction of our Azolla has been carefully worked out by Professor Campbell, of Stanford University.

The Equisetum, or horsetail of the picture, is the most common one in California. The fruiting cones are mature in March and April, when the food-making fronds are just beginning to develop. There is another form of Equisetum with a long, unbranched stem, the fruiting cones being borne at the end. As noted in the Reader, the spores of the Equisetum are hygrometric ; when dry they cling

SUPPLEMENT

together more or less by means of their long arms, and are widely scattered by the wind. When a moist place is reached, they fold their arms and rest, sometimes locking the arms around some object that serves as an anchor. Selaginella is not deemed sufficiently attractive or common to appear in the Reader ; its near relative, the Lycopodium of the greenhouses, is a very dainty plant, and its spores are easily found. A greater club-moss, known as ground pine, is used in our Eastern States for Christmas wreaths, and sometimes finds its way here. The " resurrection plant" of the Texan plains is also a kind of Lycopodium.

CHAPTER VII.

SOME EARLY FLOWERS.

The aim of this chapter is to teach the structure and function of flowers. Of course the children will need to become thoroughly familiar with the technical names of the parts of the flower, before this or succeeding chapters can be easily followed; any simple available flower can be used for drill in names. The fact that the parts of the flowers are modified leaves, is not suggested in the Reader except in the peony illustration, Nos. 1-6. If it seem best to develop this idea, the peony affords good material. Some of the sepals are very similar to foliage leaves, others are obviously the expanded petioles of similar leaves with the foliaceous part wanting, while the petals are so similar to these latter sepals that they can sometimes hardly be distinguished from them. The evolution of stamens is not so obvious; at first thought one might infer that the filament is the petiole, and the anther the blade of the leaf with its interior cells developed as pollen grains; but forms intermediate between stamens and petals, as No. 2 in the drawing, indicate that the anther rises between the petiole and the blades, the blade being usually wanting; there are traces of blade in such anthers as the violet has. Cultivated roses, indeed most double flowers, more commonly than the peony, show organs in this transition stage of petals from stamens. It is not difficult to imagine a pistil of the peony as a leaf folded on the midrib, and bearing ovules on its margins; but botanists by no means agree as to what part of the leaf is modified to become the ovary. Some hold the theory that the base of the petiole becomes ovary, the remainder of the petiole, style, when it is present, and the blade, stigma. The evolution of compound pistils is explained by various theories; so, since the botanical doctors disagree on this point, it seems best for all-round teachers to avoid putting any theory dogmatically.

When we consider the function of flowers, there are plenty of definite, universally accepted ideas to develop and emphasize. Formerly,

SUPPLEMENT

a man was persecuted as a heretic if he did not consider our world as the center of the universe, and man, the creature for whose benefit all the rest of the universe was created. So flowers existed for the gratification of man's sense of beauty, or possibly their color might indicate some magical qualities, but no one thought of the use of flowers to plants, nor of the use to flowers of their beauty, fragrance, and infinite variety of form and color. Obviously, flowers exist for the production of seed, but the significance of the opening of flowers, the fact that they expose their essential organs to the atmosphere instead of uniting sperm and egg cells within closed receptacles, could not be understood until, in our own times, the principles of nature's preference for cross fertilization had become recognized. Given the key, a study of the various means for carrying out this end, involving as it does the marvelous relations between flowers and insects, has become a fascinating pursuit to naturalists.

As remarked before, actual observation of the process of fertilization, that is of the union of the egg and sperm cells, requires skillful microscopic work, for both kinds of cells are imbedded in tissues, since they must be protected from the atmosphere. If sufficiently thin sections of pistils during the process of fertilization can be cut, it is seen that the pollen grain germinates on the stigma, sending out a tube which makes its way through the tissues of style and ovary, and through the micropyle of the ovule, so that one of the two protoplasts, or rather spermatoplasts, at the end of the tube may be brought into contact with the egg cell in the ovary; the nuclei of these two cells fuse, and it is now possible for the growth of the embryo to begin.

The conveyance of pollen from anther to stigma, on the other hand, can be easily observed, especially in our climate of almost constant sunny weather, and interest is almost sure to grow with observation. Very few flowers indeed are restricted to self-pollination; inconspicuous, greenish flowers may depend on the wind for cross pollination though insects often help in the pollination of flowers that usually escape the notice of human kind. Flowers conspicuous for color or fragrance, and those that provide honey, are, in our climate, almost always sure to receive insect help. This fact, together with many devices and adaptations, will appear as we study the individual flowers in this and in succeeding chapters.

Perhaps some preface is necessary with regard to the uses of color to flowers. That other colors than green serve to render flowers conspicuous, and so point them out to flying insects, cannot be doubted, but it must not be concluded that color in flowers can have no other use or significance. Red and kindred colors convert light into heat

CHAPTER VII

and assist in the most important of plant functions, the manufacture of carbo-hydrates. Other functions of these colors are known, and studies of the physiology of color in plants are developing truths that should make us beware of interpreting color of flowers with reference to insect visits solely. Still there seems to be no sufficient ground for discarding the old theory that white and yellow flowers are best adapted to the lower insects, white being of course also best for night pollination, and that the various shades of blue and red appeal most to guests of higher rank, bees having a decided preference for blue and violet, and humming birds for red. Unquestionably white and yellow are the most conspicuous colors against a green background, and as recent investigations seem to indicate that the lower insects have little sense of color, the greatest contrast must be most successful with them. As a matter of fact, the color of by far the greater number of flowers whose honey and pollen is accessible to short tongues, is white or yellow. In some countries tables have been carefully prepared with a view of discovering the relation between the colors of flowers and the seasons. White is the prevailing color in springtime, before the visits of higher insects can be depended upon, yellow coming next. Kerner suggests that pollen-loving insects choose yellow flowers because yellow is the usual color of pollen. Although we have no definite data in California, it will probably be generally conceded that yellow is the prevailing tone of our early flora; and there is no doubt, in Southern California at least, that scarlet flowers are most abundant during the summer months and that they have the same habitat as humming birds.

The theory that higher guests prefer blue and red, does not imply that they never share with humbler guests the hospitality of white and yellow flowers. Any one who watches the flowers and their guests in the field, knows that butterflies flit along all brightly colored flowers, though they usually choose to sip honey from deep receptacles, which are probably more common in blue and red flowers. The business-like bee clearly considers abundance of fare before color, and humming birds are frequently seen sipping honey from flowers that are not red. Still the experiments of Lubbock and others seem to prove that, other conditions being equal, bees will always choose blue; and it is in harmony with this theory that many of our blue flowers, the Brodiæa, many Gilias, some Phacelias, Salvias, sages, Pentstemons and larkspurs, for instance, keep their honey accessible to bees, but not to short-tongued guests, the theory being that these flowers have become blue through the selection of the bees. That humming birds choose scarlet, seems almost beyond question. The

SUPPLEMENT

birds often dart at scarlet ribbons or other articles, and by far the greater part of our scarlet flowers admit no other guests, except, of of course, the largest butterflies and moths. I once spent several weeks at the foot of a slope glowing with scarlet Pentstemon, while a few rods away was an equal area covered with the blue *Gilia virgata*. The Gilia furnished abundant honey, accessible, of course, to humming birds, but while I could, at almost any time of day, count a half dozen or more humming birds among the Pentstemons, I never once saw them among the Gilias. The Gilias, on the other hand, were thronged with bees.

Some good botanists advance the theory that red—excepting shades allied to violet—is distasteful to bees, and that scarlet may be invisible to them, but I have frequently seen hive bees collect pollen from scarlet Pentstemons, and carpenter bees bite through the tubes of columbine and Zauschneria for honey, while the dull red *Scrophularia Californica* is always thronged with bees, the Scrophularia, like our peony, being among our exceptional red flowers that have honey accessible to bees. The fact that insects visit consecutively flowers of the same species, can be easily verified. They do not always distinguish between similar species. I have seen bees visit indiscriminately *Gilia densiflora* and *G. virgata*. Occasionally color seems to mislead them; I have seen a bee at work regularly on the blue *Gilia multicaulis*, occasionally alight on the *Brodiœa capitata*, but usually, no matter how many flowers are intermixed, the bee selects only one species until that is exhausted.

The more common peony of California is called *Pœonia Brownii*, Dougl. That in the south, which is slightly different from the typical P. Brownii, is sometimes known as *P. Californica*, Nutt. The peony is excellent for beginning the study of flowering plants; the flowers are large and simple, and the foliage and underground parts are typical of plants of the cool, rainy season. The flowers obviously bid for cross pollination. In the field I have usually found the stigmas past maturity when dehiscence of pollen begins, so that close pollination was impossible, but the stigmas of flowers kept in the house, remain moist for several days, and an insect working in the flower might effect close pollination. If we were to interpret the color of the flower with reference to pollination alone, we should expect large flies to visit it, and the odor, too, is somewhat suggestive of carrion, but, as a matter of fact, I have never seen it visited by other guests than bees and little thrips. Really the color of the flower does not render it at all conspicuous, but the abundance of honey and pollen more than compensates for this in the estimation of the thrifty bee. If we must have a theory for the color, the most plausible one seems to be, that it helps in supplying heat, and the essential organs of the flower are said to be specially sensitive to cold. The seeds of the peony mature during the summer; I told that the fruits (follicles) explode and fling their seeds,

The buttercup, *Ranunculus Californicus*, Benth., gets its family and generic name from its supposed association with the frog pond, Ranunculus meaning tadpole. This species is abundant on moist hillsides. The peony belongs to the same family, Ranunculaceæ, because, like the buttercup, none of its parts are united. The story of the pollination of the plebian buttercup, with its indiscriminating hospitality, is rather fully told in the Reader. As in the peony, the stigmas wither early when the flowers are freely visited by insects. They do not seem to be capable of self pollination, although close pollination may be effected by insects about the second or third day. The larger insects are sure to alight in the center of a flower like the buttercup. I have seen large flies and even butterflies work as systematically as bees in gathering every drop of honey from this flower, an uncommon occurrence, I think, for usually flies and butterflies work intermittently and are not very effective agents. The little hooked style remains as part of the fruit (an akene) and helps to distribute the seeds.

Two functions of the appendages of the cluster lily, or Brodiæa, have been noted, they serve also by their contrasting color to guide the guests to the honey. The opening buds of the Brodiæa seem to me to contain even more honey than the expanded flowers; at any rate, bees often try to force them open, and every visit to a bud means cross pollination. The Brodiæa, because of early flowering, scant provision of honey, and limited range of guests, doubtless needs to be able to pollinate itself. Its fruits are capsules, which mature late. It is easy to see where new bulbs, or more properly, corms, are formed, but it is not so obvious how the upper corm retains its depth, and how the side corms get more widely separated. The key to the mystery is found when one digs up a colony of Brodiæas, that is, a larger plant surrounded by a ring of smaller ones. The little plants forming this ring about the mother plant, have sprung from the side corms of last year, and each of these corms has sent out a long and extraordinarily thick root, that by contraction is pulling it directly away from the parent corm. By means of these pulling roots, the circle of daughter plants widens each year and gets farther from the mother corm, so the Brodiæa is well disseminated and is always abundant, persisting even in grain fields where it is never allowed to mature seed.

SUPPLEMENT

The root-stock of the violet, *Viola pedunculata*, T. & G., is a very deeply buried underground stem, being sometimes eight or ten inches down. The color of this violet is not in conformity with the orthodox coloration theory; it seems to be better policy in this case for the flower to make itself conspicuous than to appeals to the aesthetic sense of the bees, at any rate this yellow violet succeeds better in California than any of its blue congeners. In countries of greater rainfall, violets often produce what are called cleistogamic flowers, that is, flowers that do not open at all, but fertilize themselves in the bud; but I have not succeeded in finding such flowers in our climate. In spite of rather infrequent insect visits, violets mature abundant seed. The seeds are usually flung from the capsule as the valves spring apart.

The shooting star, Dodecatheon, affords a sharp contrast to the peony in its methods of pollination. Providing no honey and little pollen, it must be showy to stand any chance of insect attention, and a more striking combination of colors cannot well be devised. Its device for self pollination at the last, is one practiced also by the nightshade. I have rarely seen insects visit *Dodecatheon Clevelandi* of the southern valleys, but the mountain species are rather frequently visited by bees. It should be noted that the pedicels, at first erect, recurve to allow the corolla to protect the pollen, but become erect again in fruit. The capsules contain small seeds which mature quickly. The seeds are distributed by what Kerner calls the pepper-box method, and a very effective method it is. In some species the valves of the capsule separate at the tip, leaving an opening surrounded by minute teeth; in others a small lid falls off, but in both cases the small opening allows the seeds to escape only when they are thrown out with considerable force by the swaying of the capsules in the wind. The advantages of massing flowers in clusters should be much emphasized. Small flowers scattered against a green background are invisible to insects flying at any considerable height. There is some suggestion here of the principle of coöperation, which is more fully developed in Chapter XII on "social flowers."

The calla is treated rather fully in the Reader. The difference in temperature between the interior and the exterior air can be tested with the thermometer; it may be several degrees. The pistillate flowers seem to me still receptive after the dehiscence of the staminate flowers on the same spadix, and callas do produce seed in our climate, whether from close pollination or only from cross, I have not determined. Neither have I observed the Chinese lillies out of doors sufficiently to speak of their pollination; but their growth from the bulb resting in water is a very pretty, interesting process, and can be easily watched in the schoolroom. The story of the pollination of the iris has been often told. Briefly it is as follows :—The way to the honey is marked by conspicuous beards on the reflexed sepals ; the petals are the three ascending, inflexed, floral leaves. The three innermost leaf-like organs that curve outward and downward toward the sepals, are styles ; beneath their bifid tips are small white shelves, whose under surface is stigmatic. The anthers also lie close beneath the styles and dehisce downward. Now if a guest is large enough to be of use to the flower, he must, on entering, via the honey guide on a sepal, strike his head against the stigma and immediately afterward brush it against a fresh supply of pollen from the overarching anther.

Of the other early wild flowers likely to be encountered in collecting for this lesson, those on shrubs or trees will be treated in the next chapter and its supplement, some having been already noted in Chapter V ; others will form the subjects of succeeding chapters.

The miner's lettuce, or Indian lettuce, *Claytonia perfoliata*, Donn., grows always in shady places, and is easily recognized from its succulent leaves, which are united at the bases forming little saucers, the upper pair holding the cluster of small white or pinkish flowers. In specially favorable places I have seen these flowers large enough to rival their Eastern cousins, the spring beauties ; these larger flowers were much frequented by bees, and were dependent on them for pollination, but usually the flowers are inconspicuous and pollinate themselves. The chickweed, *Stellaria media*, With., is very ubiquitous at this season in shady places, in both cultivated and untilled soil. It is a dainty little plant with its small, white, star-like flowers, which have five bifid petals and usually but three perfect stamens. Along one edge of its weak, four-angled stem is a line of hairs that are very interesting under the microscope, their function being to absorb rain and dew for the benefit of the plant.

The California four-o'clock, *Mirabilis Californica*, Gray, is a vigorous perennial, growing with the cactus on the hills of Southern California. Its showy magenta flowers are usually out in January. It is a troublesome plant to identify because what appears to be calyx is really an involucre, and the apparent corolla is calyx. It is not available for in-door study, because the flowers open only out of doors ; they open about two or three p. m., one day and close at eleven a. m. of the next. For field study these flowers are excellent; the prominent anthers and stigma lie against the lower part of the flower, the stigma always beyond the anthers, and it is easy to see whether or not the guests pollinate them.

SUPPLEMENT

If time and material admit, it may be well to collect other early-blooming members of the lily family and deduce the family characteristics. In Southern California one might find the graceful brown lily, *Fritillaria biflora*, Lindl., some of the wild onions (Alliums), the Muilla (Allium spelled backwards), a slender, onion-like plant, whose petals are white with green lines, and *Zygadenus Fremonti*, Torr., a coarse perennial with a tall panicle of greenish-white flowers and narrow, parallel-veined leaves. It will usually be best, however, to wait for more showy lilies, and defer the consideration of family traits until Chapter XIII.

CHAPTER VIII.

THE AWAKENING OF THE TREES.

Trees appeal to many people who take little notice of smaller plants, and teachers will do well to foster this natural interest. As has been already suggested in Chapters III and V, different types of trees, at least one evergreen and one deciduous, should be under continuous observation, so that their behavior at different seasons can be noted. A study of the habits and physiology of trees, especially of their relation to climate, is surely better suited to school work in our climate than the detailed study of naked twigs frequently pursued where trees are leafless or dormant half of the school year. Of course the position of buds and their plan of development, is most easily observed on leafless trees. It may be well to note that buds are situated where it is most convenient to distribute food to them, either terminally or in the axils of leaves where the conductive tissue branches off from main stem to leaf stem. Naturally there will be room for the development of only a limited number of buds each year, but when regularly developed shoots are destroyed, it is interesting to watch the activity of dormant buds; indeed, the promptness of trees to rally from accident or mutilation is little short of marvelous. The topping of shade trees furnishes an instructive illustration; nature at once covers the cut surface with new tissue from which arise numerous buds, and soon the tree is provided with a new and denser crown. We frequently see a close, and somewhat circular group of sycamores that have originally arisen as suckers from the stump or roots when the parent stem has been cut down.

Distinguishing between trees that branch excurrently and decurrently, seems to me fruitless work for children, especially as cultivated trees are rarely allowed to develop naturally; but hardly too much stress can be placed on the different characteristics of the trees adapted to a dry climate, and those which with us can survive only near streams or under irrigation. Of course, in our dry atmosphere,

SUPPLEMENT

densely umbrageous trees, such as are characteristic of humid countries like England, are not found at all, but there is a marked contrast between the Eucalyptus, live oak and pepper on the one hand, and the sycamore, maple, walnut and umbrella-tree on the other.

A fuller study of the stems of trees than has been suggested in the Reader, could doubtless be carried on with much interest and profit, especially in the upper grades. All text books of Botany treat of stem structure rather fully, and the more recent ones, such as Spalding's and Bergen's, include the physiology of the stem also. From twigs of various ages, and from pieces of firewood, one can, without the microscope, make out layers of bark, the cambium layer, wood, pith, medullary rays, and rings of annular growth. Very beautiful wood sections, prepared by R. B. Hough, Lowville, N. Y., are often included in school supplies; Mr. Hough has recently added sections of many California woods to his stock. These sections show clearly to the unaided eye, the medullary rays, rings of annual growth, and even to some extent, cellular structure. For the details of cellular structure the compound microscope is of course necessary. The outside skin, or epidermis, of stems perishes early, and the layer of green cells just beneath the corky layer also disappears in early years. The corky layer, so greatly developed in cork oak, is protective in several ways; against transpiration, changes of temperature, attacks of fungi, mechanical injuries, and so forth. It contains the same tallow-like substance, suberin, as the cuticle of mature leaves. The inner layer of bark consists of bast tissue; it contains long, tough cells, which serve for strength only, and more delicate ones, called sieve cells. The soft bast tissue plays a very important part in the nutrition of the plant, for it conducts the elaborated sap from the leaves toward the roots, the medullary rays serving to convey the sap laterally from bark to pith The woody tissue is made up of wood cells and wood ducts, or vessels, as shown in Fig. 10, Chapter II; it serves for support and for the conduction of crude sap from roots to leaves. The pith, which forms a main part of the young stem, usually disappears in time; where it exists, it conducts elaborated sap and serves to store food during the dormant period, the soft bast and medullary rays also serving the same purpose. The amount of food stored in stems above ground is very considerable. In the trunks of old, yet vigorous trees, a considerable part of the wood, as well as the pith, is dead; the outer part of the bark, too, is constantly wearing away, sometimes falling in plates or shreds; but in every active tree there must be living wood to conduct crude sap from roots to leaves, and living bark to distribute elaborated sap. Between these living cylinders of wood and bark is the cambium layer, the layer of new cells, which during the growing season multiply very rapidly, becoming new wood and new bark. It is the delicacy of these new cells in spring time that makes it possible to slip the cylinder of bark off the wood of willow and elder twigs. Early in the growing season the wood cells have comparatively thin walls and long diameters, later the tissue is much more compact; this gives rise to the rings of annual growth.

The coverings of buds, the disposition of leaves in the bud, and the devices for protecting the young leaves as the buds develop into shoots, are interesting topics. The nature of the wrappings can be made out by the older pupils. That the scales are often modified leaves, is clearly shown in developing walnut buds, there being many intermediate forms between the simple scale and the compound leaf. Many of our trees use stipules for bud wrappings, not only the fig, India rubber and magnolia, but the apple, cherry and their kind, and the oak, alder and others. In the sycamore the collar-like structure is a pair of united stipules, but in this case the stipules become a permanent part of the foliage. Leaves in the bud have their conductive tissue very prominent, but the parenchyma is not well developed, and there is little or no cuticle on the epidermal cells. As the leaves throw off their wrappings the delicate parenchyma cells are in special need of protection against over-transpiration, and the same devices that serve this purpose, also protect against cold and moisture. It is not against frost chiefly that protection is needed; leaves need to retain at night, as far as possible, the heat acquired during the day, since growth goes on night and day. The various protective devices are easily seen. Often the soft tissue is plaited, folded or crumpled in, the veins only being exposed. Usually the young leaves assume a vertical position; besides this, there are coverings of varnish, or woolly or silky hairs. Sometimes there are membranous stipules that grow and continue to screen the young leaves until they have attained considerable size, as in the magnolia and the India rubber. The use of the red color to young leaves has already been noticed in Chapters III and V. This color, and the vertical position, seem to be all the protection the young leaves of the Eucalyptus and the pepper require.

The willow in mind when this chapter was written, was *Salix lasiolepis*, Benth., an early flowering species common throughout the state; other species might differ in detail. The pistil of the willow, when the fruit is mature, splits into its two carpels. The fluff attached to the seeds floats them on water as well as in the air. Another aid

SUPPLEMENT

to the distribution of willows, is their habit of striking root so readily that severed twigs are very likely to become new plants. The poplar, or cottonwood, Populus, is nearly related to the willow, as is apparent on examining catkins and fruit. There are several stamens on a disk in each staminate flower, and a pistil may be composed of three or four instead of two carpels. The catkins come early, before the leaves. This tree is not very common in Southern California, but is well worth observation where it does occur.

The Sycamore, *Platanus racemosa*, Tourn., alder, *Alnus rhombifolia*, Nutt., walnut, *Juglans Californica*, Wats., or the oak, Quercus, may be studied as typical of wind-pollinated plants. The flowers of the walnut are the easiest to study on account of their size, but it would be unfortunate to miss any of the others that can be observed by the children in the field. The sycamore is one of the most beautiful of our cañon trees; it sometimes attains an imposing size; one is known to be nine feet in diameter. In the spherical flower clusters each pistil or each stamen is a flower. The pollen is protected by the connectives of the anthers, which fit closely together until the pollen is mature and the weather dry; then, by the shrinking of this tissue, the anthers are sufficiently separated to allow little clouds of pollen to escape with every breeze. The staminate catkins of the alder mature and fall in December or January, and soon the pistillate clusters become little cone-like fruits; the leaves come usually in March. The delicate beauty of these straight, slender trees bursting into leaf, is one of the delights of our spring time. Each walnut flower has a calyx, there are many stamens in each staminate flower, and the pistillate flowers have inferior ovaries and minute petals.

Nearly all wind pollinated plants with staminate catkins have some provision for the temporary deposition of pollen; the cottonwood has the same device as the walnut; the anthers of pines and firs deposit pollen on the back of neighboring scales; the anthers of the sycamore have their connective tissue expanded at the ends like nail heads, and the pollen is lodged here, and so on. The pollen lying in loose little heaps is very easily lifted by the wind, and, as a matter of fact, pollen is best distributed, not by a strong wind, but by gentle rising currents of hot air. About one-tenth of all flowering plants are pollinated by the wind; some of their characteristics are, small, inconspicuous flower, large stigmas and abundance of pollen, easily accessible to the wind. Bees frequently collect pollen from walnuts and other trees of this class, but they are not likely to be of any service.

The oak has quite as interesting leaf buds, flowers and fruits as the

CHAPTER VIII

other trees. Birds as well as rodents collect acorns, but it is doubtful whether they aid in propagating the oak, since woodpeckers deposit the acorns in holes made in the bark of trees, and blue jays put them in hollows in trees, into the pith of broken elder branches and like places; whether the numbers lost and scattered by the birds is considerable I do not know. The elder, *Sambucus glauca*, Nutt., is often of sufficient size to be counted with trees. It is nearly evergreen in our climate, and is rather attractive, with its fresh foliage and large clusters of white flowers. Its flowers are pollinated by insects; the fleshy fruits will be considered later. The Spanish Californians use the leaves medicinally; the fruit, also, is esteemed.

The maple, *Acer macrophyllum*, Pursh, and the bay, or laurel, *Umbellularia Californica*, Nutt., of our cañons, are not generally accessible for class work; neither, in Southern California, are any of our native conifers, except such cultivated species as the Monterey pine, *Pinus insignis*, Dougl., the Monterey cypress, *Cupressus microcarpa*, Hook., and the "big tree," *Sequoia gigantea*, Decaisne. These, or any other cultivated conifers, can be watched as they push out new leaves, shed their clouds of pollen and mature their cones. Most of them are in flower in January or February. The staminate flowers of the pine are most impressive, but the fertile clusters and young cones of the cypress are generally more accessible. The staminate catkins of Coniferæ consist merely of scales bearing pollen sacs, two each in the pine, and from three to five each in the Sequoia and cypress. The fertile flowers, which, after fertilization, develop into cones, consist of scales bearing naked ovules, two each in the pine, more in the cypress. The illustration shows staminate and fertile flowers of an introduced pine and cypress, and the development of a cypress cone from the tiny green cluster on the branch to the woody cone that has opened and allowed the seeds to escape. This development is easily seen on the trees; the fleshy scales close over the ovules soon after pollination, grow larger and more woody, and ultimately separate to allow the seeds to escape. Conifers are very deliberate in all of their habits; many of them, like the Monterey pine and cypress, require two years to ripen their seeds, and some retain their seeds even five years. Our native conifers and trees generally, will be further referred to in Chapters XIII and XIV. Some facts as to the size and age of trees will be given in the Supplement to Chapter XIII. It is well to cultivate the habit of measuring the finest trees available.

The awakening of the deciduous trees of our orchards is full of interest from the time that the swelling buds impart wondrous tints to

SUPPLEMENT

the trees in mass, up to the development of the fruit. Bud protection can be easily discerned; the flowers and fruit should be studied as they appear. The stone fruits, apricots, peaches, plums, etc., are classified as drupes; apples, pears and quinces, as pomes. These two groups are usually included in the rose family. If the teacher looks forward to some work in classification, the children should make and preserve drawings showing the parts of these flowers, also of strawberry, raspberry or blackberry flowers when they are accessible. Except for occasional variations, such as usually occur in cultivated plants, these flowers all have five sepals, five petals and many stamens, all situated on the calyx. The pistils vary; the ovary of the drupe group is superior and consists of one carpel, the pome group has five inferior ovaries imbedded in the receptacle, while the so-called berries have many small pistils situated on a convex receptacle. The pollination of the flowers should be noted; they usually seem capable of self pollination, still they provide much honey, and obviously advertise this by color and fragrance. Experiments seem to show that many trees of this family fruit better when cross pollination is effected. It is said that there are varieties of pears that must be pollinated not only from another tree, but from a different variety. Our wild blackberries, and some kinds of cultivated strawberries, have frequently only staminate or pistillate flowers, and so require cross pollination.

The subject of the development and distribution of fleshy fruits should not be missed later on. By comparing flower and young fruit it is seen that the ovary wall of a drupe becomes the edible part. The ovary walls of pomes are simply the five chartaceous pods that contain the seeds; botanists do not agree as to whether the fleshy part we eat is altogether receptacle, or both receptacle and calyx. Obviously the juicy part of the strawberry is the greatly enlarged receptacle, what are commonly called seeds being entire pistils. But the receptacle of the blackberry remains white and tough; the juicy part of the fruit is the mass of ovaries, each ovary having become a drupe. The receptacle of the raspberry remains on the bush when we pick the united drupes. So these berries are not berries at all in the botanical sense, a berry being a several seeded ovary, pulpy throughout, such as the elderberry, cranberry, currants, grapes, oranges, tomatoes, etc.

Until their seeds are mature all of the fleshy fruits are inconspicuous and have some bitter or possibly poisonous qualities; then by chemical changes the juices become sweet and well flavored. Obviously the taste, together with the color and fragrance, are devices for tempting animals to eat the fruits. Sometimes the seeds are not swallowed but are thrown away, more often they pass through the alimentary canal. Many experiments have been made to determine to what extent seeds of fleshy fruits are injured by the digestive fluids; only the most general results can be stated here. Seeds swallowed by birds are retained only a short time, from one-half to three hours, and a small percentage only are injured; on the other hand, more than half of those swallowed by animals are rendered incapable of germination. That some poorly protected seeds like oats and barley, escape injury is evident to everyone who has used fresh stable manure as a fertilizer. If fleshy fruits are not eaten, but simply fall to the ground, the decaying fleshy part anchors the seeds to the soil.

To return to the topic in hand, the awakening of the trees: Young figs appear with first leaf buds, early in the year. The pistillate flowers in the interior can be readily discerned. The pollination of some kinds of figs is described by Kerner in Vol. II, page 159-160, of his "Natural History of Plants." Briefly, the story of our common species is this :—One tree bears edible figs, containing only pistillate flowers; another tree of the same species bears what are called caprifigs, having near the orifice staminate flowers, but farther up, gall flowers, i. e., undeveloped pistillate flowers in which the egg of a wasp has been placed. The female wasps in escaping from the caprifig become covered with pollen; they pass at once to younger figs to lay their eggs, and if they chance to select a tree bearing figs with developed pistillate flowers, they of course pollinate them and the seeds become fertile; the wasp's egg rarely develops in these figs. If the wasp enters caprifigs, her eggs will develop in the imperfect pistillate flowers, but the caprifigs fall off before becoming juicy. Kerner states that while, in some districts, the caprifigs are cultivated and hung on the fertile trees to ensure pollination, this is quite a waste of labor, for the flavor of the figs is not in the least improved by fertilization; and, since figs are now propagated by cuttings, it matters not at all that the seed is infertile.

It is unsafe to make any general statements about the habits of some of our introduced plants, especially if they are irrigated. Of Eucalyptus and Acacia, for instance, we have many species that come from very different climates. The point insisted upon is that children be encouraged to watch continuously some tree or trees in the vicinity of the school building, so that their observations can be tested and discussed. A few additional facts about trees will be brought out in Chapter XIV.

SUPPLEMENT

There is no sharp line between trees and shrubs, sometimes habitat determines whether a given species is to become a tree or a shrub. In their habits, shrubs, as well as trees, present much of interest, and there is every reason why children should be helped to a growing acquaintance with the most common and most attractive shrubs of their district. Some deciduous shrubs that awaken very early, have been already noted in Chapter V, the poison oak, some species of Ribes, and *Ceanothus crassifolius*. The poison oak, *Rhus diversiloba*, T. and G., is very often erect, but against a steep bank or a tree it frequently asserts its power to climb by rootlets. It blossoms in early spring, and the bees frequent its greenish white flowers. *Rhus trilobata*, Nutt., somewhat resembles the poison oak, but it is usually lower and has smaller, greener and more numerous leaves. The twigs of this Rhus are used by the Indians for their finest basket work, and its fruits are eaten. *Rhus integrifolia*, B. & H., and R. *ovata*, Wats., of the south, are evergreen, and form dense thickets. They are sometimes called Indian lemonade bushes on account of the acid outer coat of the fruits, and sometimes mahogany trees because of their hard wood. Their pretty white or rose-colored flowers appear in December or January. They provide honey, and are cross-pollinated by bees, some plants being staminate, others pistillate. Another Rhus of Southern California, *R. laurina*, Nutt., is abundant near the sea and on the coast islands. Its large, evergreen leaves are very aromatic and have often much red coloring matter, suggesting the common name sumac; it blooms later than the others.

In the south the wild lilac, or Ceanothus, already described, is soon followed by *C. divaricatus*, Nutt., and other species. In the north *C. thyrsiflorus*, Esch., attains tree-like proportions. The flowers of all these species provide honey, some of them lavishly, and the honey is much appreciated by the bees. The stamens, which at first are distant from the pistils, later on bend over and place their anthers against the stigmas.

The yellow-flowered currant, *Ribes tenuiflorum*, Lindl., blooms in January or February, just in time to be pollinated by the earliest carpenter bees. The gooseberries noted in Chapter V, are soon followed by the more common *R. speciosum*, Pursh., with tubular, scarlet flowers. These flowers are usually too long for any but hummingbirds and the largest bees to get honey in a legitimate way; but smaller bees sometimes steal the honey by biting through the calyx tube. This Ribes has beautiful, glossy leaves and is densely clothed with spines, so that it is gaining favor with gardeners as a hedge shrub.

CHAPTER VIII

There are many other shrubs common in the foothills. The most common "scrub" oak is *Quercus dumosa*, Nutt.; in habit it is similar to other live-oaks. The California "holly" is *Heteromeles arbutifolia*, Roem.; it is only its red berries, maturing in early winter, that suggest the name holly; its leaves do not at all resemble the English holly, and its flowers are white and come in midsummer. There is a wild cherry, *Prunus ilicifolia*, Walp., with pretty, evergreen, spiny-tipped leaves that is really like the true holly. This is a very common shrub on dry hillsides; under more favorable conditions it becomes a handsome tree. Another wild cherry, *P. demissa*, Walp., has deciduous leaves and fruit with more pulp; in Southern California this cherry is limited to the mountains. Both cherries have fragrant, white flowers in spring time, and are much visited by bees. Perhaps our most useful shrub is *Rhamnus Californica*, Esch., since from its bark the highly prized medicine, cascara sagrada, is obtained. The common name of this shrub is coffee plant, since its dark red fruits, with their two great seeds, resemble coffee berries. The true wild honeysuckle, *Lonicera* (or Caprifolium), with pale flowers and perfoliate leaves, blooms in late spring or in summer; its near relative, the *Symphoricarpus*, or snow-berry, matures its white fruits in autumn. The botanical name for the manzanita is *Arctostaphylos;* for the madroñe, *Arbutus,* and for the mountain mahogany, *Cercocarpus betulæfolius*, Greene; all of these bloom in early spring time.

There are some showy shrubs that are not so widely distributed. The tree poppy will be noticed in the next chapter; *Fremontia Californica*, Torr., is a very striking shrub of the drier foothills of the Sierras. There are shrubs or small trees belonging to two genera of the family Malvaceæ; the *Lavatera assurgentiflora*, Kell., or tree mallow, has large, rose-colored flowers, and nutritive, mucilaginous leaves, of which grazing animals are very fond; this malva is a native of the coast islands, and has become rather common about seaboard towns. Several species of Malvastrum in the southern foothills are shrubs, one of them, *M. Davidsoni*, Robinson, is sometimes fifteen feet high. Like other plants of arid regions, they are of slender habit, and their gray-green foliage forms a fitting background for the exquisite, delicate-pink flowers. The introduced shrubs, the castor oil plant, and the tree tobacco, are referred, perhaps unjustly, to the chapter on weeds, and a number of the lower shrubby plants that bloom in summer are described in Chapter XV. The woody climbers, the wild blackberry and the grape, should be noted in field work with the shrubs and trees.

CHAPTER IX.

SOME SPRING FLOWERS.

With children who have already considerable acquaintance with our wild flowers, it will be well to emphasize the relationship of the plants of this chapter. The poppy family has marked traits; two of these, the fact that the sepals fall off as the flower expands, and that there are twice as many petals as sepals, will serve to place all the members we are likely to meet. The family as a whole is a very showy one. The Spanish name "copo de oro," cup of gold, is the best one for the *Eschscholtzia Californica*, Cham. There are some kindred species difficult to distinguish from this, but this is by far the most common, and it varies greatly in size and color. In rich soil in the Mohave Desert, the size of this poppy and its richness of color are magnificent; the inhabitants speak of these poppy fields, not as golden, but as blood-red, and in the distance this is the color effect. The normal time for blooming is in early spring, but poppy flowers may be found at almost any time of the year. If a first crop is destroyed in grain fields, a second may spring up in summer, or the earliest autumn rains may rouse the plants with strong perennial roots.

The pollination of the poppy is typical of flowers that furnish only pollen. The color renders it very conspicuous, the amount of pollen is considerable, and it is carefully hoarded. The cream cup, tree poppy, and the two large white species furnish even greater supplies of pollen. The sleepy habit of the plants is, of course, only a device for protecting pollen. Kerner states that the older flowers simply furl each petal like a tent over its own pollen, but I have rarely found this true of uncultivated Echscholtzias, and it is not a common habit in our gardens; the petals usually roll up together as in the bud. Kerner states also that the stigmas of the smaller flowers are short enough to be in contact with the anthers, while in larger flowers they are above the anthers, a condition which by no means always holds with the poppies at home; neither do I find that they fling their fruits quite away from the receptacle; all of which only shows that plants behave differently under different conditions. Our poppies can sometimes pollinate themselves when they are open; they are almost sure to do so as they close, and the insects that seek shelter in them must also effect close pollination; still much pollen is carried from flower to flower, and some botanists think them quite infertile to their own pollen. It is probable, though it has not been proved, that when flowers receive both their own and another flower's pollen, the foreign pollen is prepotent.

The flame-colored poppy, *Meconopsis (Papaver) heterophyllum*, to be fully appreciated, must be seen lifting its nodding buds and satiny flowers from some grassy slope. The flowers in the south are not more than an inch in diameter, in the north they are sometimes two inches.

The tree poppy, *Dendromecon rigidum*, Benth., rivals the Eshscholtzia in splendor; the flowers are not so large nor so deeply colored, but the shrubs, which blossom most profusely, are sometimes fifteen feet high. In Southern California the tree poppy chooses sandy washes, and while the plants are sometimes washed away by floods they are also distributed by this means; occasionally one gets stranded within the city limits of Los Angeles. The acme of showiness is attained by the two white species, *Argemone platyceras*, var. *hispida*, Prain, and *Romneya Coulteri*, Harv. Both have great, crinkled, white petals and numerous yellow anthers, and the bumble bees fairly wallow in their pollen. The Argemone is widely distributed in the southwest. It flowers in summer and loves hot, sandy washes. I have seen plants six feet high standing alone in the sand. The Romneya is often known as the Matilija poppy, because it is abundant in a cañon of that name in Ventura County; but it is found in many other cañons in Southern California, and is gaining rapidly in favor as a cultivated plant. The cream cup, *Platystemon Californicus*, Benth., is of more modest dimensions, but it is justly a great favorite; it often quite takes possession of waste or semi-cultivated soil and produces myriads of creamy white flowers. In exposed places the flowers may have a liberal dash of red coloring matter. One member of this family, common in damp cañons, *Platystigma denticulatum*, Greene, is remarkable for its delicate beauty; it has little, star-like, white flowers and is likely to be mistaken for a lily unless one discovers the three green sepals of the bud.

The mustard family, Cruciferæ, has also marked characteristics, as noted in the Reader. Most Cruciferæ have dehiscent fruits, and when the two valves separate at maturity, a transparent partition remains, framed by the placenta or seed-bearing portion of the fruit. Some of the plants of this family that children would be likely to find, are enumerated in Chapter XIV of the Reader. The shepherd's purse, *Capsella Bursa-pastoris*, Medic., and the pepper grass, *Lepidium nitidum*, Nutt., are common lawn or wayside weeds that begin flowering in January. Their tiny white flowers continue to develop at the end of the cluster, long after the lower part has ripened its fruits. The fruits are tiny; those of the shepherd's purse are obcordate, the others are round. The botanical name for the watercress is *Nasturtium officinale*, Tourn.; for sweet alyssum, *Lobularia* (formerly Alyssum) *maritima*, Desv., and for the wild radish that overruns neglected places, *Raphanus sativus*, Linn. All three of these can be found in bloom almost any month. The wild turnip, *Brassica campestris*, Linn., usually known as mustard, is very common in the outskirts of civilization, blooming earlier than the black mustard; it has smooth leaves, the stem leaves clasping at the base. The field mustard, or black mustard, *Brassica nigra*, Koch., is further treated in the Reader in Chapter XVI. The lace-pod, *Thysanocarpus curvipes*, Hook., has minute white flowers, but exceedingly pretty fruits; the fruits are not dehiscent, that is, the ovary does not open and discharge the seed, but it develops a dainty, lace-like margin, called a wing, which aids in seed distribution. The plant is slender, but bears many long racemes of pale-green, or pinkish lace-pods; it blooms in spring time in rather sheltered places.

The western wall-flower, *Erysimum asperum*, DC., is a stiff, rough, hardy plant, with large, fragrant, orange-colored flowers; it begins to bloom in mid-winter, and its long racemes last for many weeks. Having more handsome flowers than most Cruciferæ, the wall-flower can afford to be more exclusive, and its sepals cohere so closely that it has practically a tube about half an inch deep. The stigma is mature when the bud begins to expand, and as the expansion is very deliberate and the honey abundant, the pollination of any flower is likely to occur before the dehiscence of its own pollen begins. All Cruciferæ mature stigmas before shedding pollen, and as honey is usually fairly abundant and is frequently accessible to any insect, there must be a considerable amount of cross pollination. So far as I have observed, there has always been some provision for self pollination later on, and insects often effect close as well as cross pollination. There are many other common Cruciferæ, but it would be impossible to distinguish them clearly without tedious description and technical terms.

Œnothera bistorta, Nutt., is a southern species, but there are several other species with similar flowers and habits. The beach species is very handsome, and serves well as a type of seashore or desert vegetation, having very long roots that lie near the surface ready to utilize every fog and dew, and foliage well protected from the intense light and heat. When children can do field work on the beach they should be encouraged to measure roots and find different sorts of protective coverings. *Œnothera Californica*, Wats., sometimes grows very luxuriantly in warm, sandy soil. It is a very handsome plant with its silver-grey foliage and large, white, flowers, which finally take on rosy tints. The flowers open about five p. m., and usually close before noon on the next day, but on the sand dunes of the Mohave Desert, I have seen them remain open for thirty-six hours. They are obviously adapted for pollination by large night moths; the honey is sometimes an inch deep in the long tubes, and both the sticky stigmas and the long anthers covered with a webby mass of coarse pollen, are directly in the way of the entering guest. The yellow evening primrose, either *Œnothera biennis*, Linn., or some of its varieties, is more commonly met than the white species; it has rather coarse foliage and, while evidently adapted to night pollination, it is frequently found with flowers open throughout the day. Godetias, Clarkia and Zauschneria are taken up in Chapter XV.

The five sympetalous families or orders grouped together in the Reader are very nearly related; the first four are well represented in California. It is easy to memorize both the characteristics that are common to all, and the character of the pistil, which determines the family. The order Polemoniaceæ, to which the genus Gilia belongs, has always a three-celled ovary and three stigmas; Hydrophyllaceæ, including the genera Nemophila and Phacelia, has a two-celled ovary and a style more or less two cleft; Boraginaceæ is the heliotrope and forget-me-not family, and has always a four-parted ovary, but one style and one stigma; Solanaceæ, the potato and nightshade family, has but one ovary, which, in many genera becomes a berry in fruit.

Gilias are particularly abundant, and are very easily recognized as Gilias, though the specific name may be difficult to determine. The kinds pictured in the Reader are very abundant in Southern California. *Gilia multicaulis*, Benth., varies extremely in size, according to its habitat. In poor, dry soil it may be only a few inches high and consist of a single stem bearing two or three pale flowers; in more favorable places it may be a foot high, with clusters so many-flowered that unless one has learned to know it from its fragrance, it is difficult to distinguish it from *G. achilleæfolia*, Benth. *G. capitata*, Dougl., more common in the north, is another Gilia of similar appear-

SUPPLEMENT

ance and habits, so is *G. tricolor*, Benth., except that its flowers are more showy, the three colors being pale violet, purple and gold. I have seen this last Gilia cover acres of so-called desert land in the Mohave, the mass of color being a striking feature of the landscape. In this locality the high winds preclude insect visits much of the time, but I have seen very active carpenter bees making the most of a brief lull in the wind; besides, the anthers bend to the center of the flower so that the pollen must fall down the throat and tube and so come in contact with the stigmas as the corolla falls There is a pretty, golden Gilia, *G. aurea*, Nutt., that is also able to flourish in hot, sandy places in the south; its flowers are niggardly in providing honey, and must often have to pollinate themselves.

G. dianthoides, Endl., one of most exquisite of California wild flowers, does literally carpet many a sunny spot in the southern sections. It, too, is variable in habit; in grassy places it may send up branches several inches high, when unshaded it may spread out in densely flowering mats, or again may consist of a single stem lifting a flower an inch or so above the earth. A species very nearly related to this, *G. dichotoma*, Benth., is named evening snow, its white flowers opening at four or five p. m., and closing at daybreak; like most night pollinated flowers it has a heavy odor. *G. micrantha*, Steud., and some allied species with white or delicately tinted flowers, do not tell the story of their pollination so clearly. The flowers are abundant in late spring, or during the summer in the mountains; the corolla tubes are extremely slender and sometimes an inch or more long, so that only large butterflies or moths can reach the honey; in the valleys the flowers usually close at night, but in the mountains I have found them remaining open; the stamens in the mountain species vary in length but the styles are constant, so they are not, as has been thought, dimorphic; I have reached no satisfactory conclusion about them, nor about the *G. Californica*. This latter species, growing in masses among the chapparal, is a peerless little shrub, and it is an anomaly indeed, if the showiness of the flowers be of no use to the plant.

Of the Nemophilas, *N. aurita*, Lindl., presents by far the most legible story, but the others are very justly popular favorites. *N. insignis*, Dougl., is perhaps the only species to which the name baby blue eyes should be applied; certainly it has few rivals in blueness. According to color theories we should expect to find the flowers in close alliance with the bees; they have five pairs of appendages arranged to contain honey, but those I have examined have contained very little, and the amount of pollen was small. The flowers close

CHAPTER IX

early and remain closed in unfavorable weather, and the bees seem to seek them mainly for shelter. Honey is hardly perceptible in the exquisite, pale blue *N. Menziesii*, H. & A., and I have rarely seen it visited ; but both of these species have larger flowers in the north and may have quite a different story of pollination. The flowers probably pollinate themselves when they close.

There are about forty species of Phacelias in California, and many of them are nearly allied to *P. tanacetifolia*, Benth. The leaves of this group are much dissected, being bi-pinnate or tri-pinnate, and the flowers are nearly always of some shade of blue ; the name wild heliotrope has arisen because of their color and form of inflorescence, heliotrope really belonging to the family Boraginaceæ. The flowers of *P. tanacetifolia* seem always well provided with honey, which is well protected and excluded from all but useful guests. The anthers face upward during dehiscence, and the stigmas also, when they mature, are held where they are sure to be struck by the guest. I know no native flower that gets a greater share of attention from the bees than this species, and the group as a whole is very successful. There are other Phacelias with entire leaves ; one of these, *P. circinata*, Jacq., a rough, hoary plant, is very widely distributed ; it bears white flowers in late spring and summer.

A considerable number of Phacelias have notched, viscid leaves ; four of these species with large, handsome flowers are common in the south. *P. grandiflora*, Gray, has mottled blue and white, rotate flowers nearly two inches in diameter, but one rarely handles the plant a second time, because it leaves a stain like iron rust ; it is also poisonous to some people. *P. viscida*, too, has rotate flowers ; they are nearly an inch in diameter and are often a very intense blue. *P. Whitlavia*, Gray, and *P. Parryi*, Torr., grow in rich soil along river bottoms or banks ; the flowers of both are a rich royal purple, those of the former are bell-shaped and are commonly called Canterbury bells ; the flowers of the other species are rotate with five white honey guides near the centre. Both of these flowers seem to depend on the splendor of their attire to attract guests, for they supply very little honey ; large bees occasionally visit *P. Whitlavia* for honey, and hive bees often collect pollen from the conspicuous white anthers of *P. Parryi*.

Besides Nemophilas, Phacelias, and the Ellisia mentioned in the Reader, two other genera of Hydrophyllaceæ contribute widely distributed plants :—*Emmenanthe penduliflora*, Benth., is common on dry hillsides; its most common name is whispering bells, from the fact that its yellow, bell-shaped corollas, instead of falling off, become dry

SUPPLEMENT

and rustling, and cling until the fruit is mature. *Eriodictyon glutinosum*, Benth., is the yerba santa, whose medicinal value, discovered by the Indians, is still appreciated by the medical profession. This species is generally confined to the mountains of Southern California, but the species *E. tomentosum*, Benth., is common in lower altitudes. The Eriodictyons are shrubs with thick leaves and bluish flowers in scorpioid clusters; the leaves of the yerba santa are resinous, and the flowers are frequently white. *E. tomentosum* is worthily named, for the "tomentum," or dense hairs, clothes not the leaves only, but stems, calyxes and even the violet corollas. The flowers seem to be pollinated by bees.

California has a wild heliotrope, *Heliotropium Curassavicum*, Linn., a low, succulent plant, common near the sea and in moist alkali soil the world over; it blooms in summer, having small white or bluish-white flowers. Of the Borraginaceæ that bloom in the spring time, those with the yellow flowers belong to the genus Amsinckia, and the white ones to different genera not agreed upon by botanists. There are supposed to be several species of Amsinckia in Southern California, but the one so common as a roadside weed is *A. spectabilis*, F. & M. The ability of this plant to withstand drought, is probably its strongest point, but it is generally well equipped. Its flowers are uncommonly large for this family; they have abundance of honey and prominent honey guides; sometimes they are slightly irregular. Any one studying their pollination will be impressed with the variability in the length of stamens and pistils; some plants have flowers with anthers well above the stigmas, others have stigmas always above the anthers. If it were not for many intermediate forms we should consider these flowers dimorphic, and possibly they are developing in this direction. True dimorphism is explained in any book that treats of pollination, the English primrose being the favorite illustration. If an insect come from a plant with high anthers to one with high stigmas and low anthers, it cross pollinates this flower, and pollen from the low anthers adheres to its tongue; passing to flowers of the other form with this supply of pollen, the insect cross pollinates the low stigmas and gets a fresh supply of pollen from high anthers, and so on. There are also flowers with well marked trimorphism; I have found no clear case of either among native California flowers. The bristly calyx of *Amsinckia spectabilis*, which gives it its name of "woolly breeches," is persistent and aids in seed distribution.

As stated above, great confusion still prevails over the nomenclature of our wild white forget-me-nots; three genera, Krynitzkia, Plagiobothrys, and Pectocarya are represented in the vicinity of Los

CHAPTER IX

Angeles. The one in the illustration in the Reader is perhaps the most attractive, coming as it does against the delicate background of early vegetation; the children call it pop-corn flower. Many others of the white forget-me-nots are rough or prickly.

The family Solanaceæ includes the nightshade, potato, tomato, Chili pepper, egg plant, tobacco, tree tobacco, ground cherry, "Jimson" weed, and many showy cultivated plants, such as Petunia, Stramonium, and several climbers with large, handsome flowers. The common, weedy nightshade of the south is *Solanum Douglasii*, Dunal., sometimes considered a variety of *S. nigrum*. The fragrant, blue nightshade is *S. Xanti*, Gray; on Catalina Island the variety *Wallacei* of this species has extremely large, handsome flowers. I have yet to learn of any case of poisoning from *S. Douglasii*; children frequently eat the berries with impunity. I have known one instance of poisoning from the handling of S. Xanti. The pollination of both species is given in the Reader. The tree tobacco, *Nicotiana glauca*, Tourn , and the California "Jimson" weed, *Datura meteloides*, D C., will receive attention in Chapter XVI. Our native tobaccos, *N. Bigelovii*, Wats., and other species, are disagreeable herbs hardly common enough to merit much attention. The ground cherry, *Physalis æquata*, Jacq., is a rather pretty introduced weed not uncommon in loamy fields. It has nodding, yellow flowers and fruits enclosed in pretty inflated calyxes.

Our wild morning-glory, Convolvulus, is always interesting in the field, but is of little use for in-door study because the flowers collapse so quickly. Cultivated morning-glories are easily and quickly grown in our climate, and their twining will be watched with much interest; it is readily seen that the tips actually revolve, and that they will revolve in but one direction. The most common wild Convolvulus of Southern California is *C. occidentalis*, Gray. Its flowers provide some honey, but it is so hidden by a combination of ovary, stamens and corolla tube that I have never yet seen an insect get it except by biting through the corolla. Bees sometimes collect pollen from the anthers that form a dome over the forked stigma, and in doing this they must effect close as well as cross pollination. This morning-glory has a strong perennial root, and is somewhat persistent even in cultivated fields, but the Convolvulus that is proving so troublesome in vineyards and orchards in some parts of the state, is a virile European weed, *C. arvensis*, L. The dodder, Cuscuta, belongs to the order Convolvulaceæ, but will be considered in Chapter XV.

The nodding white Mariposa, *Calochortus albus*, Dougl., is found in the foot-hills and mountains throughout the state, and one can hardly

SUPPLEMENT

imagine a more exquisite flower. The yellow globe tulip, *C. pulchellus*, is common north of Monterey County. *C. albus* is frequently much larger than the one in the illustration, the flowers being often an inch in diameter. I have seen the flowers visited by flies and beetles only; the flowers seem capable of self pollination. Of the cup shaped Mariposas, perhaps the one that is most fittingly named the butterfly tulip, is the *C. venustus*, Benth. It is rather widely distributed; I have found it in cañons bordering the San Fernando Valley, and abundantly at Newhall. The corollas vary from white to violet, rose or cream, and there are three blotches of color on each petal, the upper one usually red. In form the flowers are exquisite beyond description. *C. luteus*, var. *oculatus*, Wats., is said to be equally beautiful. The most common Mariposa about Los Angeles and other adjoining coast towns, is the one illustrated, *C. Catalinæ*, Wats., which usually shades from white to lilac or rose, and has crimson spots. *C. splendens*, Dougl., is common in the south; it is clear lilac throughout, and occasionally has no honey glands at all. There is a sturdy yellow Mariposa, *C. clavatus*, common in southern cañons, and *C. luteus*, a smaller plant with yellow flowers is also found. The very hairy, purple-blotched *C. Weedii*, var. *purpurascens*, Wats., is rather abundant in the south, but it blooms in June and July and so misses the attention it deserves. *C. Nuttallii*, of the high mountains is found in all shades from cream-white to deep purple. The flame-colored Mariposa, *C. Kennedyi*, Porter, is common in the Mohave Desert. As implied in the Reader, it seems probable that these Mariposas must often fail of pollination. The stigmas open very late, and even then it is easy for an insect to get honey or pollen without striking them. I have never found honey very abundant, and the hairs about the glands must be a veritable thicket to a small insect; at any rate I have seen insects come for pollen much more frequently than for honey. The bulbs become rather deeply buried; and some species produce bulblets freely in the axils of the lower leaves.

Perhaps mention should be made of a section of Calochorti not represented in the south, but common in the north, the group popularly known as the star tulip; the flowers resemble those of the globe group, but they are erect or nearly so, and are from bell-shaped to star-shaped rather than globular; they are of various tints, and are sometimes known as "mouse ears," or "pussy ears;" *C. Maweanus*, Leich., is a type.

In Southern California the plants considered in this chapter will usually be abundant in March or early in April. The next three chapters should follow as promptly as possible, especially if the two

CHAPTER IX

on classification, Chapters XIII and XIV, are to be taken up. Even in this way, many common and interesting flowers will be missed. Some larkspurs are sure to be found long before they are discussed in Chapter XV. If the children collect these, they should be taught the name and encouraged to find out the parts of the flowers, note the insect guests, etc. Often a member of the Portulaca family, either *Calandrinia Menziesii*, T. & G., or *C. elegans*, Spach, will be very abundant along waysides or in grain fields. The plant has fleshy leaves and numerous bright, rose colored or magenta flowers. The amount of pollen compensates for the lack of honey, and the great number of little, black seeds should by all means be noted, since their abundance probably accounts for the persistence of the plant in cultivated soil.

Have the children at least note the fact that the malva weed, *Malva parviflora*, Linn., has flowers, and that the "cheeses" are the ovaries. If there is time to compare these flowers with those of the wild hollyhock, Sidalcea, or with the shrubby Malvas, the family characteristics, especially the column of united stamens, numerous pistils and mucilaginous juice, can be noted. The flowers of some members of the rose family will have to be noted here, if at all; the wild blackberry, *Rubus ursinus*, for instance, also some strawberry-like flowers, Potentillas. In collecting in well shaded places some herbaceous saxifrages will probably be encountered, *Tellima affinis* perhaps, or *Saxifraga reflexa*. Fleshy leaf rosettes belonging to the "live-for-ever" family, Crassulaceæ, will be encountered, but they usually bloom rather late, and will be mentioned in Chapter XV, along with the Cacti. The large pea family, Leguminosæ, and the geranium family come in the next chapter, and a chapter on irregular sympetalous flowers follows.

Nearly all members of these families now in bloom will last for some weeks, though probably not that pretty little Labiate, the skullcap, *Scutellaria tuberosa*, Benth. But some of the most common Umbelliferæ, and some attractive Compositæ are likely to be missed entirely. Let the children at least note the existence of these flowers and give them common names when possible. The family name can be easily developed. There is one Umbellifer, a Sanicula, with flowers in low, yellow clusters, appearing as early as January; this can be called yellow mats or meadow sanicle; there is also an early, purple-flowered sanicle. The tidy tips, *Layia platyglossa*, Gray, and *L. elegans*, with their yellow, white-tipped rays are among the prettiest and best known of California Compositæ. In Southern California there are two other early Compositæ about the size and general appearance of the tidy tips, but lacking the white tips; their botanical

SUPPLEMENT

names are *Leptosyne Douglasii*, DC., and *Senecio Californicus*, DC. The Leptosyne has a long, leafless flower stem and a double involucre, a few linear bracts forming the outer circle, the inner consisting of broader, overlapping bracts. The genus Senecio is commonly known as the groundsel; this species has a leafy stem and the akenes have always a capillary pappus. The most common of all the early daisy-like Compositæ, the fragrant one that literally carpets hundreds of acres, is *Bæria gracilis;* it is sometimes called fly flower, sometimes sunshine or golden fields. The grass family will not come regularly till Chapter XIII, but the children can be led to discover before this the fact that grasses have flowers, and can become acquainted with at least the wild oats, wild barley or fox-tail, and—if it is common in their vicinity—the Bermuda grass. The so-called blue-eyed grass, *Sisyrinchium bellum*, Wats.,—see Fig. 6; also Fig. 2, No. 2—is, of course, not a grass at all, but its pretty blue flowers should be noted and named.

CHAPTER X.

PLANTS WITH MECHANICAL GENIUS.

The many mechanical devices of the flowers, fruits and leaves of these two families, Leguminosæ and Geraniaceæ, seem to me to justify the title in the Reader. There are so many species of lupines that their identification is difficult. They are usually grouped into perennials and annuals. The most common perennial in the vicinity of Los Angeles, is *Lupinus albifrons*, Benth., an almost shrubby plant with silken leaves. It blooms the year round, but more profusely in spring time. The flowers of the long raceme are usually pale blue, but in the San Fernando valley I have seen them pink; their fragrance renders the species unmistakable. *L. Chamissonis*, Esch., is the common shrubby lupine of the beaches. The long roots of these and of kindred species make them valuable in confining sand dunes. A notable instance of this was the transformation of once shifting sand hills into the beautiful Golden Gate Park of San Francisco. By sowing, first barley, which grows very rapidly, then the lupines, the soil was prepared for shrubs and trees and other vegetation. Another perennial lupine common in the foothills of Southern California, is *L. formosus*, var. *Bridgesii*, Greene; it resembles *L. albifrons*, but is not so shrubby. The perennial that grows so luxuriantly along streams is *L. cytisoides*, Agardh. I have seen this lupine grow in thickets, the leaves tropical in size, and the long clusters of rose-colored, fragrant flowers above one's head. The flowers of this species belong to the larger bees; hive bees are not heavy enough to pump out the pollen, but they sometimes cling to the keel and wings, and scratch out the pollen. The annual lupine of the illustration is *L. sparsiflorus*, Benth.; other common annuals are *L. affinis*, Agardh., an early succulent plant; *L. truncatus*, Nutt., with slender, truncated leaflets; *L. hirsutissimus*, Benth., the very rough lupine of dry habitats; and *L. micranthus*, Dougl., with very tiny flowers. *L. densiflorus*, Benth., is common in many parts of the state; it has succulent foliage, handsome white, pale yellow or pinkish flowers, and pods with but two seeds.

SUPPLEMENT

Children should become thoroughly familiar with the lupine as a type of papilionaceous, or butterfly flowers, for the main group of the order **Leguminosæ** is very well represented in our fields and gardens, and its flowers, fruits and leaves have such pronounced characteristics, that the kinship of the plants is easily apparent. The piston apparatus of the lupine flowers is in working order as soon as the banner is erected, and is efficient until the upper edges of the keel separate. The stigma is not the brushy part of the style, but is merely the tiny tip that is surrounded by bristles. Perhaps the bristles serve to keep off the flower's own pollen until after the first insect visit, but ultimately there is sure to be close pollination. Whether all species are fertile to their own pollen I do not know. Some of the most showy lupines rarely mature fruit on the upper part of the raceme, but *L. micranthus*, which I have never seen visited, matures abundance of fruit. Naturally, lupines with the most showy and fragrant flower clusters, receive the most insect attention; the most successful one I have seen is *L. confertus*, Kellog, in the San Bernardino Mountains, and the bee that paid thirty-five calls per minute was the Nevada bumble-bee.

The alfalfa, *Medicago sativa*, Linn., wherever it is grown, is too much appreciated to need extended exposition; its roots not unfrequently attain a depth of eight or ten feet, and they sometimes go much deeper. It is hoped that the story of its pollination is made clear in the Reader; with the flowers in hand it is easily seen; self pollination must always occur, and the flowers are said to be fertile to their own pollen. Butterflies in great numbers are often seen fluttering over alfalfa fields, but I have not had opportunity to determine whether or not they usually explode the flowers; bees certainly steal the honey quite as often as they get it in the direct way. The coiled fruits lack the great advantage of hooked appendages, but they are very readily blown about in the dust. The flowers of bur-clover, *Medicago denticulata*, Willd., in spite of their minuteness are eagerly sought by bees.

California has a long list of native clovers, many more species than the Atlantic States, but they do not grow abundantly, and the fragrant clover fields that are such a feature of our North-Eastern States are unknown in California, at least in the southern half of the state. Our most striking native clover, *Trifolium fucatum*, Lindl., is a succulent plant with large heads of pink or pale rose-colored flowers, the individual flowers being perhaps an inch long. This handsome clover is found in the foothills of both mountain ranges, and is sometimes abundant enough to be valuable as a pasture plant. In Southern

California, *Trifolium tridentatum*, Lindl., and *T. involucratum*, Willd., are rather common in loamy soil ; the former, found in dryer soil, has narrow, acute leaflets and small heads of purple and white flowers ; the latter, found where there is more moisture, has more luxuriant foliage, and flower clusters that resemble the cultivated *T. pratense*, L., of the North Atlantic States. These species, however, vary extremely in different localities, and are sometimes difficult to distinguish from each other and from kindred species. They have abundant honey, exclude short-tongued guests, and seem able to pollinate themselves, but whether they are fertile to their own pollen I do not know. Some of the most valuable cultivated clovers will not produce seed without cross fertilization, and as our Eastern farmers clear away the forests that harbor the bumble-bees, they have to import clover seed. Australian farmers have imported the bees with the clover. Our so-called wild peas and vetches come under two genera, Lathyrus and Vicia. The *Lathyrus* of Southern California is a vigorous climber that blooms often as early as January, and continues in bloom all spring. The flowers are so variable that it is difficult to decide whether there are several species or several varieties of one species (vestitus) ; most commonly they are white or cream color with veins of violet or rose, but they are often entirely rose-colored, violet or nearly purple, and one sort has exceedingly handsome crimson flowers. The flowers of *Vicia Americana*, var. *truncata*, Brewer, are often confused with the others ; they are more slender, and perhaps more bluish in color, but a sure way to distinguish them is by the style ; in both genera there is a brush of hairs just below the stigma or tip of the style ; in the Vicias this brush extends around the style, in the Lathyrus it is on the inner surface only. In both genera this tiny brush serves an important end. The anthers shed their pollen early into the keel and over the brush and stigma, which, however, is not yet mature. The wings and keel are firmly united by means of pouch and socket, as in the alfalfa ; and, as in alfalfa, the honey is accessible through two openings at the base of the upper, or free stamen. At the base of the flower the banner fits closely over the rest, and only strong insects can force an entrance to the honey ; hive bees can, with considerable scrambling, succeed with the Vicia, but not with the Lathyrus ; but the larger native bees, particularly bumble-bees, are frequent guests with both species. As the lower part of the flower is forced down, the style brush sweeps out some pollen against the insect's body, and the friction of the guest against the stigma is said to render it receptive to the foreign pollen the insect brings ; such is the elasticity of the tissue involved, that the stamens and style

SUPPLEMENT

almost invariably resume their secure **position** in the keel after a visit; that is, each flower may repeatedly **receive foreign pollen on its** stigma, and, in return, furnish its guests with fresh supplies. Much of this remarkable process can be seen by imitating the pressure of the bees on the flower. The climbing habits of these plants are similar to those of our cultivated **sweet** peas, to which they are nearly related.

The "rattleweed" **genus**, Astragalus, is a very large one, and about one hundred **and fifty** of the five or six hundred species, are natives of Western **America. In** our fertile valleys there are **several species,** the most **common being** *A. leucopsis*, Gray; but the numerous desert and **mountain** species, with their various adaptations to hard conditions, are the most interesting. The inflation of the pods aids in seed **distribution. Doubtless some species are poisonous to** sheep and cattle, but it is very doubtful whether the so-called "locöing" observed in this section, is due to anything else than the presence of bezoars in the stomachs of the animals.

The genus Lotus—the Hosackia of the older floras—is particularly well represented in California. The species *L. glaber*, Fig. 11, is widely distributed. In the mountains it is called deer weed, because the deer feed on it. In dry regions I have seen it with the leaves reduced almost to the vanishing point, and **generally, because of its** numerous green stems which act **as foliage, it deserves its name,** California **broom. It is one of the most attractive chapparal plants, blooming to some extent the year round, and very profusely in spring and early summer.** It is an important bee plant, and so are many of **the annual species common in** the state. The pretty little *L. strigosus*, Greene, with its fine, rather rigid leaves, and *L. Wrangelianus*, F. & M., with succulent leaves, have few rivals in the attentions of **hive** bees; they have devices for pollination similar to those of the **Lathyrus.** There are other, more handsome species that are widely **distributed, and are worthy of** individual attention. Several species of Lotus have the same device for seed distribution as the Lupines, **that** is, the valves coil back, so expelling the seeds forcibly. Other Leguminosæ will be referred to in Chapters XIV and XVI.

Few of our native Geraniaceæ are conspicuous, and the kinship of the introduced members of the family, the **filaree**, Pelargonium, the so-called nasturtium, and the Oxalis is not easily made apparent to **children, but all these genera** have mechanical contrivances, some of **them very striking.** Perhaps the story of the filaree is told in sufficient detail in the Reader; the young plants are referred to in Chapter V, and the weed-like qualities of the plant are discussed in Chapter XVI. **Its** flowers have a primitive and wasteful method of pollination; they are

CHAPTER X

frequented by all sorts of insects, many of them quite useless to the plant; but the bee appreciates their hospitality, and I have seen bees lick up the honey still remaining in flowers whose petals had fallen. The fact that the fruits bore themselves into a soft substratum is made more real when children find them imbedded in fleshy fruits or flowers, in strawberries or callas, for instance. We have a native Erodium and several species of geranium with fruits very similar to the "clocks;" the flowers of these species, except some in the mountains, are small and not easily observed. The flowers of the native Oxalis, too, on account of their "sleepy" habits and scarcity of honey, are not satisfactory for passing observation. There is a variety of *Oxalis corniculata*, a common garden weed, that has explosive fruits, one of the inner layers of the ovary wall consisting of tissue in a state of tension that is quite apparent to the touch. The garden balsam, Impatiens, which has such strikingly explosive fruits, is not so commonly cultivated in California as in the Eastern States.

The so-called geraniums of cultivation are mostly Pelargoniums from South Africa. They are particularly adapted to our climate for many reasons that can be easily thought out; one device, the absorption of moisture by their hairs, was noted in Supplement to Chapter III. Their methods for pollination are striking and easily discerned; the long duration of the clusters should be noted. Only newly opened flowers supply pollen; the anthers drop off usually before the five stigmas begin to unfold. Honey is kept in a "concealed spur," that is, in a tube united with the flower stem, the entrance to which is between the two upper petals; the three lower petals form a platform, the flowers being slightly irregular, and the guest alighting on these petals, or rising from below as humming birds do, must reach over anthers or stigmas to get the honey; hence exactly the same place on the guest's body that, in newly opened flowers, strikes the anthers will, in older flowers, be rubbed against the stigmas. The tube varies in depth in different species, but in our common red, white and pink "geraniums" it is usually quite too deep for bees. Butterfly collectors tell us that they capture many of their finest night moths among the geraniums, and one has but to walk along the residence streets of a city to see the humming birds at work on the scarlet varieties. Other common cultivated species, the ivy-geranium, or the "Lady Washington" for example, may have shorter tubes with honey accessible to bees, and some of them have very striking honey guides. The Pelargoniums produce abundant seed in our climate, the fruits having precisely the same devices as the filaree.

The "nasturtium" of cultivation, which is really a Tropæolium

SUPPLEMENT

(the true Nasturtium is the water cress), belongs to the Geranium family, but has not so many mechanical devices. It is, however, usually available for out-of-door observation in California, and it is very attractive and interesting, particularly in its method of climbing, and its pollination. The honey in this case, too, is in a spur, which is not concealed by union with the petiole. The flowers, being solitary, need to be larger and to have their sepals, as well as their petals, brightly colored. The honey guides are prominent, the stamens and styles behave precisely like those of the larkspur, see Chapter XV. Humming birds frequent the flowers and are the most efficient visitors. Bees and flies sometimes come for pollen; perhaps the larger bees get honey also.

The leaf movements so common in Leguminosæ and so striking in Oxalis, were noted in Chapter V, but should be taken up again. It should be observed especially, that these leaves assume their various "sleeping" positions, all of them practically vertical, during dry winds or a prolonged drought; even the filaree erects and folds its leaves to some extent. Obviously, the same device that restricts radiation of heat, also restricts transpiration of moisture.

CHAPTER XI.

PLANTS OF HIGH RANK.

The application of the term rank to plants is not perhaps in strict accord with the theories of modern biologists concerning species. Certainly it is not easy to define what constitutes the highest development in plants; but in order to emphasize, in this and in following chapters the differentiation of parts and the division of labor existing in certain groups of plants, the old term " higher plants " is retained. The fact that there are so many different systems of classification, none claiming perfection, and that there is no one plant or plant family that has universal recognition as the highest, is noted in the next chapter. The bilabiate families, Scrophulariaceæ and Labiatæ are particularly well represented in California in the late spring and summer, and it is very easy to recognize their members; they have always irregular corollas, two or four stamens, and one style; the ovary of the Scrophulariaceæ is always two-celled, the ovary of the Labiatæ, like that of Borraginaceæ, separates into four nutlets. Two families might possibly be confused with them; the family Verbenaceæ, which is commonly represented by a rough, weedy plant with small, crowded, violet flowers, *Verbena prostrata*, R. Br., and Orobanchaceæ, the cancer root family, which consists of root parasites, and hence is without chlorophyll, the entire plants being a dull, brownish yellow or purple.

Mimulus glutinosus, Wendl., illustrates well the capacity of flowers for color variation. The most common color is buff or salmon, but, in the mountains I have seen masses of flowers that were hardly more than cream colored, while at Catalina Island I have found the flowers red only. Frequently in the foot-hills one finds, growing in close proximity, plants of this Mimulus with buff flowers, others with red ones, and others in intermediate shades. Bees are quite debarred from the honey of these flowers; they cannot even bite through the flower, because of the tough, viscid calyx that closely invests the corolla tube. There are small insects that can enter the tube bodily, but, as the flower has an open throat, they are not likely to strike the essential organs. Small beetles that have been feeding on pollen in older flowers are sure to effect cross pollination as they enter expanding buds, since the open stigma always guards the entrance to buds. I have but once seen humming birds visit the yellow flowers, and have had no opportunity to watch for night moths; the red flowers at Catalina are rather frequently visited by humming birds; that their red color is due to the selection of the birds is, of course, only theory.

Mimulus luteus, Linn., has the throat much narrower, but I have seen small insects enter without touching the stigmas. This plant is common everywhere along running water, and often grows in masses, but I have never, except in the mountains, seen it visited by bees; in the valleys it frequently fails to mature seed. There is another Mimulus common in Southern California, that has large, handsome, yellow flowers; this is *M. brevipes*, Benth., an extremely viscid annual, found on sandy banks and hillsides; this, too, secretes very little honey, and I have never seen it visited. Our one hospitable member of this genus seems to be *M. cardinalis*, Dougl., a large and exceedingly brilliant flower, found along our streams in summer time; this flower supplies honey abundantly and keeps its anthers and stigma where only humming birds can be of any service. There are, in warm soils, several low annual species which have rather large, rich, magenta flowers, but I have had little opportunity to watch them. The smaller, yellow species that abound along mountain streams and meadows, seem to be quite as parsimonius as their larger yellow cousins, but they have generally some device for narrowing the throat.

The *Collinsia bicolor*, Benth., is one of the prettiest of our late spring annuals. In some localities it is known as "innocence," and I have heard children call it Chinese temple, probably because of the many-storied effect of its flower-cluster. The corolla is a remarkable imitation of the papilionaceous type, the two united upper petals corresponding to banner, the folded lower one to keel, and the lateral to wings. There is a rudimentary fifth stamen that serves as a honey gland. Honey is abundant, and the method of pollination is easily seen; hive, and other large bees, visit the flowers frequently. Several other species of Collinsia are not rare. The owl's clover, or pink painter's brush, and the scarlet painter's brush, belong to different but nearly related genera. The former, *Orthocarpus purpurascens*, Benth., is very common, sometimes becoming a weed in grain fields; its common name, clover, is suggested by the color, not by the structure of the flowers. The other plant in the illustration is *Castilleia parviflora*, Bong., but several other species resemble it closely; the hoary *C. foliolosa*, H. & A., of dry hillsides, is the only one easily identified. In both these genera, the bracts and calyx tips contribute much to the general showiness, the slender corollas having their two upper petals quite united, and the three lower ones very much reduced. The flowers in both cases have abundant honey; I have rarely seen day insects avail themselves of the honey in the Orthocarpus, but I have marked numbers of unvisited stigmas at night, and have found them quite visibly pollinated in the morning. The Castilleias, I have seen pollinated only by humming birds, though sometimes the flowers are short enough for bees. Both plants produce abundant seed.

The Pentstemon that is called the scarlet bugler, is *P. centranthifolius*, Benth.; the violet one of the illustration is *P. heterophyllus*, Lindl., its flowers being very similar to *P. spectabilis*, Thurb. Some later Pentstemons will be referred to in Chapter XV. I have frequently seen the violet Pentstemons visited by large native bees, and by regal Masaria wasps that take siestas in the flowers. Several other genera of the family Scrophulariaceæ are rather commonly encountered. There are several mulleins, Verbascums, imported weeds, that may be found in waste lands. These woolly plants send up wand-like clusters several feet high; the flowers are but slightly irregular and, unlike other members of this order, have five stamens, with anthers; the filaments are densely bearded. The Veronicas, or speedwells are found only in moist places; the small white or blue flowers have but two stamens. *Linaria Canadensis*, Dum., the blue toad flax, sometimes grows in masses in soil that has been cultivated; the plants may be a foot or more high, but are exceedingly slender; each blue flower has a long, slender spur, and, like the snapdragon, has the throat closed. The botanical name for snapdragon is Antirrhinum, but none of our native species have sufficiently large flowers to be popularly recognized as snapdragons. *A. Coulterianum*, Benth., common in sandy soil in the south, has wand-like stems with branches that serve as tendrils; its white flowers form a spike, and, while not very showy, are well visited by bees that are strong enough to open the corollas. Climbing species of Antirrhinum with blue flowers are not rare. *Scrophularia Californica*, Cham., is a rank, weedy-looking plant with reddish-brown, chubby little flowers. This genus has such a patent story of cross pollination that it is usually selected for an illustration in text books. The flowers are nearly globular, but have a wide opening between the upper and lower lips; between the two

SUPPLEMENT

upper petals is the rudimentary fifth stamen reduced to a scale; the four perfect stamens lie against the lower lip, but at first are entirely within the corolla. When the flower opens, the style with its stigmatic tip looks over the edge of the lower lip; here it is sure to be struck by the bees or wasps that come for honey. After pollination, the style curves closely over the lip, and puts the withering stigma out of the way, then, two by two, the anthers pop up and stand exactly where the stigma stood. Honey is particularly abundant in these flowers; bees come in throngs, and I have seen even humming birds visit them; so these flowers are a striking illustration of the fact that abundance of refreshment counts more, even with the highest guests, than imposing size or pleasing color. There are two genera obviously nearly related to Castilleia and Orthocarpus; they are Cordylanthus and Pedicularis. The flowers of the former are found in summer or in dry habitats and are not striking, but the latter blooms early in damp woods, and the most common species is so striking, with its reddish stems and leaves and its handsome crimson clusters, that children call it Indian warrior. It is found but rarely in Southern California.

The family Labiatæ has, besides the characteristics already mentioned, square stems and opposite leaves; the flowers are solitary or in whorls in the axils of leaves; when they are densely clustered, the effect is of a succession of heads through which the stem passes. Nearly all Labiatæ are aromatic, and several genera are called mints, Mentha, Monardella and Micromeria for instance; *Micromeria Douglasii*, Benth., is the yerba buena so prized as a medicine by the Spanish Californians; *Monardella lanceolata*, Gray, or "pennyroyal," is a handsome-flowered mint, rather common in the mountains; its flowers are well patronized by bees and butterflies. Several species of Stachys, or hedge nettle, are rather common plants of weedy habit; *Stachys bullata*, Benth., is found in moist places; the flowers have an obvious story of pollination. The Scutellariæ are called skullcaps because of their peculiar helmet-shaped calyxes. *S. tuberosa* is only a few inches high, but it has vigorous, tuberous roots, and large purple flowers that come early, as noted in the Supplement to Chapter IX. Another Scutellaria, with similar flowers, is common in the mountains in summer, and is pollinated by large bees. The hoarhound is classed with weeds in Chapter XVI, and so is *Trichostema lanceolatum*, Benth., the disagreeably scented blue-curls. *Trichostema lanatum*, Benth., the woolly blue-curls, is a pleasantly scented shrub, with unique handsome flowers; it is rather common on gravelly hillsides in Southern California.

CHAPTER XI

The genus Salvia is properly the sage genus, but our native plants called sages, belong to the allied genus Audibertia. Both genera have very peculiar stamens. The tissue joining the two cells of an anther is called the connective; in these flowers the connective is usually developed so that it separates, rather than connects, the two cells. In the Salvias both anther cells contain pollen, and the connective rests on the filament like a see-saw board on its support. In the genus Audibertia but one anther cell remains, and the connective is simply joined at or near the other end to the filament, so that the stamen appears to be a spliced filament bearing a single anther cell at the top. In the little chia, *Salvia Columbariæ*, Benth., the value of this curious mechanism appears; as the guest, usually a large bee, thrusts his head into the flower for honey, he strikes it against the lower anther cell, the connective swings on the filament and the other anther cell strikes the bee's back, so that he is well dusted with pollen to carry to older flowers, whose stigmas are mature. The big chia, *S. carduacea*, Benth., has its filaments so short as almost to escape notice; the two parts of the connective are rigid, and are easily mistaken for filaments; in fact the essential organs in this flower do not seem well adjusted, and pollination is very much a matter of chance.

Watching the pollination of the black sage, *Audibertia stachyoides*, Benth., and the white sage, *A. polystachya*, Benth., is delightful field work. The process is fully described in the Reader. Another sage common on hillsides in some parts of Southern California is *A. nivea*, Benth. In this species the stamens and style are longer than in the black sage, and stand about midway between upper and lower lip, so that the bees generally use them as a perch while they are gathering honey; this brings pollen-covered anthers in younger flowers, and mature stigmas in older ones, against the bee's body. A handsome mountain sage will be noticed in Chapter XV. The botanical name for the crimson sage is *A. grandiflora*, Benth.

In the vicinity of Los Angeles, I frequently find very interesting hybrid forms, crosses between *A. polystachya* and *A. stachyoides*. The whole subject of hybridization is full of interest. Of course, the production of hybrids among cultivated plants is extremely common. It has been practiced by Chinese and Japanese gardeners from the beginning of historic times. It is estimated that about 6,000 kinds of roses alone have been produced in this way. Wild hybrids are much more common than is generally supposed. About 1,000 new hybrids were reported in Europe alone in forty years. Hybrids exist among mosses and ferns as well as among flowering plants. Kerner devotes many pages to this topic, and states, as an indisputable conclusion, that

SUPPLEMENT

hybridization does give rise to new species. A hybrid may be quite as fertile as the parent species, or even more so, and among the many hybrids that arise, some are likely to be better adapted to changing climatic conditions, changes in insect life, etc., than are the parent plants. Kerner claims that the full significance of sex in plants can be explained only in this way; that is, that the union of the oöplasm and spermatoplasm from different species can give rise to new species better adapted to new conditions.

To return to the sages and other Labiatæ; the calyxes persist and enclose the four seed-like fruits; sometimes the calyx teeth attach themselves to passing objects, and so become efficient means of seed distribution, as every one who has encountered the hoarhound knows. But generally the calyxes are not easily broken off; most persistent calyxes serve as do capsules, simply to hold seeds until they are scattered by hard winds. Kerner states that in many cases the teeth are pressed down by passing objects, or against neighboring twigs in a wind, and that in the rebound the fruits are shot out with considerable force.

CHAPTER XII.

SOCIAL FLOWERS.

Members of the family Compositæ are easily recognized; the flowers are always in a dense head surrounded by an involucre; with few exceptions, there are five united anthers and a two-cleft style; the ovary is always inferior and the fruit an akene. Identification of species in a family of ten thousand is, of course, difficult; some of the primary divisions of the order are based upon characteristics seen clearly only under the microscope. So the identification of Compositæ is usually undertaken only by professional botanists, and the laity do well to recognize the principle of division of labor, and ask specialists to name these plants. But since a botanist may not always be accessible, and because this group contains many of our most handsome and interesting plants, native and cultivated, a brief description of some of the most common native Compositæ is here given.

The group Tubulifloræ includes all the Compositæ having tubular flowers, whether or not they have ray flowers also. The Aster tribe is one of this group, and includes many common orders, the greater number of them summer and autumn-blooming plants. Some of them are referred to in the Supplement to Chapter III. Not all of this tribe resemble asters in general appearance. The gum plant or resin weed, Grindelia, belongs here; it is not so common in the south, but farther north the species *G. cuneifolia*, Nutt., abounds in salt marshes and is collected in great quantities for the manufacture of a medicine of the same name, a medicine that, like the cascara, was discovered by the Indians, and is now in general use. The heads have yellow rays, and are rather large and showy; the involucres are covered with a milky resin that is the valuable product of the plant. *Heterotheca grandiflora*, Nutt., is a very common tar-weed of the south; it is a tall, stiff, coarse plant, resinous, ill scented and usually dust laden; the flowers are of medium size, and have numerous yellow rays; the fruits have a rusty pappus.

SUPPLEMENT

Asters are found all over the state, but neither they nor their near relative, the golden rod, are so common a feature of the summer and autumn flora as in the Atlantic States. They belong to three genera difficult to distinguish, Corethrogyne, Aster and Erigeron. They have both ray and disc flowers and a capillary pappus; the disc flowers are commonly yellow, the ray flowers are usually blue, violet or lavender, but they may be white or rose colored. In the southern valleys the common summer and autumn species are scattered plants, rather stiff in habit, but farther north on the beaches, and in moist places in the higher mountains throughout the state, there are some graceful and exceedingly beautiful Asters. *Erigeron Philadelphicus*, Linn., is a common weed in wet places, and has in spring time rather large and very pretty flower heads with white fringe-like rays. *E. Canadensis*, Linn., is another common weed; it flourishes late in autumn, and in land that has been cultivated, it is sometimes six feet high; its flowers are small and greenish white, but because of its numerous, slender, green leaves and branches and its fluffy fruits, it is by no means an unattractive plant.

The California golden rod, *Solidago Californica*, Nutt., is found throughout the state and is especially abundant in the mountains; its heads are small, but they are massed in fairly large clusters, not, however, at all comparable with the great plume-like clusters of some Eastern species. *S. occidentalis*, Nutt., grows along streams and has small clusters. Aplopappus and Bigelovia, as noted in Chapter III, are genera specially adapted to the arid interior regions; nearly all our species are low and shrubby, with narrow, rigid and often resinous leaves; the heads of yellow flowers are usually slender with rays wanting or not conspicuous. Our one genus that contains shrubs, the Baccharis, is also one of the Aster tribe. The tallest species, *B. viminea*, DC., called the flowering or mock-willow, grows along streams and is sometimes twenty feet high; other species grow along streams or even on coast hills, and usually have numerous red nutgalls on the leaves. They bloom in late summer and autumn; the heads on some plants are staminate, on others pistillate; none have ray flowers; the pappus is very copious and silky.

The Gnaphaliums, or everlasting flowers, belong to another tribe. Although some species bloom in spring time, the plants are so well adapted to hard conditions, either aridity or cold, that they are included in Chapter III. The Eidelweiss of the Alps is a Gnaphalium. The individual flowers are so very slender that it is not easy to distinguish their parts; there are no ray flowers, and only a few central flowers in each head are perfect, the outer ones being pistillate only.

CHAPTER XII

The fruits of this genus, like all the others we have considered so far, have a capillary pappus. The Ambrosia tribe, in spite of its mellifluous name, contains some ugly weeds; the genus Ambrosia includes the rag-weed, so troublesome in Eastern States, but on our coast the allied genus Franseria is more common, and equally weedy. These weeds have some redeeming features; the foliage is always pleasing, the leaves being pinnately divided, often very delicately. On the beach, or in sandy wastes, the plants are grey, sometimes silky, but in rich cultivated soil they have, like the fleabane, green foliage when everything else is brown or gray. Another genus of this tribe is Xanthium, the cocklebur, the impersonation of total depravity among plants. The flowers of these genera might not be recognized as belonging to Compositæ; the heads of staminate flowers are clustered, but are inconspicuous; on the same plant are the pistillate flowers only one or two in a head, surrounded by involucres that are covered with spines, the hooked spines of the cocklebur being specially vicious.

Many of our common Compositæ are included in the sunflower tribe and two or three allied tribes that the amateur finds difficult to distinguish. The common sunflower, *Helianthus annuus*, Linn., supposed to be the ancestor of the cultivated sunflower, is referred to in Chapter XVI as taking possession of cultivated land during the summer and autumn; it grows also along streams and in warm, loose soil generally, and may be found in flower almost any month. *Encelia Californica*, Nutt., of the south, is often mistaken for the sunflower; it is a handsome, strongly scented perennial, growing in masses and blooming very fully in spring time; the heads are somewhat smaller than sunflowers, usually about two and a half inches in diameter; the plants are bushy, being woody at the base like a marguerite, but they are not so coarse and rough as a sunflower. Their method of pollination does not, I think, differ from that of the sunflower.

Of the Compositæ mentioned in the Supplement to Chapter IX, the tidy-tips, *Layia platyglossa*, Gray, is nearly related to the sunflower and has the same methods; another, *Layia glandulosa*, H. & A., and its variety *rosea*, with ray flowers white and rosy respectively, are very widely distributed in sandy soil, and are successful in securing insect patronage; they can be readily recognized by any one familiar with tidy-tips. *Leptosyne Douglasii*, DC., belongs to the sunflower tribe and *Bæria gracilis*, Gray, to a very nearly allied tribe; the groundsel, *Senecio Californica*, DC., belongs to a group separated from these mainly because of its capillary pappus. Another groundsel

SUPPLEMENT

which has inconspicuous flowers, is an introduced European weed; and the handsome, shrubby, autumn-blooming *Senecio Douglasii*, is described in the Supplement to Chapter III. There are other Bærias difficult of identification, and two very large flowered species of Leptosyne are occasionally met on the sea coast or desert. The Leptosynes can be recognised by the characteristic double involucre described in Chapter IX, Supplement.

The genus Bidens, including the bur-marigold, *B. chrysanthemoides*, Michx., and the Spanish needle, *B. pilosa*, Linn., is another genus belonging to the sunflower tribe, and the large flowers of the former species, so abundant along streams in autumn, much resemble sunflowers. The akenes of both species are tipped with from two to four barbed awns, which are most efficient in distributing the seed. The species *B. pilosa* is an introduced weed common in waste places in the south; neither foliage nor flower is attractive, and the fruits are a great affliction to pedestrians.

The two tar-weed genera, Madia and Hemizonia, also belong to this tribe. They are taken up in the Reader in Chapter XVI, but as some of them bloom in May, it may be best to consider the genera here. The name Hemizonia was suggested by the shape of the involucre, which, in both genera, is hemispherical. The bracts of the involucre, in both genera, infold the akenes of the ray flowers. *Madia sativa*, Molina, blooming in July and August, is probably the most obnoxious of all this obnoxious tribe; it is unattractive and ill scented, and its resinous secretion mingled with dust ruins clothing; but this same secretion, besides protecting the plant in several ways, fastens the bracts containing ripened fruits to passing objects. Another Madia, *M. elegans*, Don., blooming in May and June, is fragrant, and has handsome flower heads that close during the heat of the day; the yellow ray flowers are often marked with brown or dark red at the base, suggesting the name wild coreopsis. These and other Madias are much more common northward than in the south; so are the Hemizonias generally, excepting *H. fasciculata* var. *ramosissima*, Gray, which begins blooming in June and becomes a common feature of summer vegetation. This plant is low but much branched; it has sparse foliage, and very numerous small flower heads, though with five broad, yellow ray flowers each. The species *H. pungens*, T. & G., has leaves and bracts tipped with spines. *H. luzulæfolia*, DC., is the most common species in the more northern harvest fields; it has a spicy odor and rather attractive white or pale yellow flower heads of medium size. John Muir devotes a page of his "Mountains of California" to the charms of another Hemizonia, which blooms in October; he speaks of the numerous flower heads—sometimes three thousand on a single plant—and of their richness in **texture and color**. There is a Hemizonia, *H. tenella*, Gray., often collected in June on southern hillsides, that is at first glance easily mistaken for a white Gilia; its three to five prominent ray flowers have the corollas very deeply three-cleft.

Nearly all the Compositæ we have been considering have both ray and disc flowers; but the genus Chænactis belonging to the same tribe as Bæria, and hence nearly related to the sunflower tribe, is quite destitute of ligulate flowers, although its heads are handsome and of good size; the outer tubular flowers are large and usually simulate ray flowers. The flowers are generally woolly, and the plants are adapted to higher altitudes or dry regions, very handsome species being found in the Mohave Desert; the heads may be yellow, white or rose colored; the most common species about Los Angeles, *C.glabriuscula*, DC., has yellow flowers blooming in company with the tidy-tips. *Eriophyllum confertiflorum*, Dougl., the golden yarrow, is also one of this tribe; the plants are not more than two feet high, but are woody and woolly, and are generally adapted to dry hillsides and summer suns; the small heads of deep yellow flowers are massed in large flat-topped clusters similar to the true yarrow. The true yarrow, *Achillea millefolium*, Linn., is put in another tribe, because of its involucres of papery or scale-like bracts; the clusters of small, white heads are large and showy; the leaves are very finely divided, and form rosettes close to the ground that are so dense and green, in spite of dry soil, that the plant is sometimes used in lawns. The yarrow is highly scented, so are most other genera in this tribe, such as the chrysanthemum, tansy, Anthemis, or Chamomile (*Anthemis cotula* being the Mayweed), Artemisia, the sage brush or wormwood, and *Cotula coronopifolia*, Linn., the common weed of wet grounds, whose heads the children call brass buttons. Some members of this tribe have heads with ray flowers, but several have disc flowers only, the Cotula and Artemisia for instance; the flowers of the latter genus are very inconspicuous and are wind pollinated.

The thistle tribe has tubular flowers only. The corollas are deeply slashed and the style branches are united nearly or quite to the tip. The true thistle, Carduus, or Cnicus, has an exceedingly downy pappus, each bristle being a long, slender **plume**. The large handsome thistle most common in the south, is *Cnicus occidentalis*, Gray; its spine-tipped leaves and bracts are clothed with long white wool and cobwebby hairs; the flower heads are about two inches long and nearly

SUPPLEMENT

as broad. The yellow star thistle, *Centaurea melitensis*, Linn., will be considered in Chapter XVI.

The sub-order, Liguliflorœ, has heads composed of ligulate flowers only; it is also characterized by its bitter, milky juice. The corollas are five-toothed; our common species have generally a downy pappus, and flower heads that remain open only under the most favorable conditions. Chicory, cultivated as a substitute for coffee, has escaped from cultivation in some parts of the state; its pretty blue flower heads are sometimes called bachelor's buttons. Lettuce and salsify belong here, and the introduced dandelion of our lawns and streets; also the sow-thistle, Sonchus, considered in Chapter XVI. The most handsome native flowers of this group belong to the genus Malacothrix. *M. saxatilis*, var. *tenuifolia*, is pictured and described in Chapter III. There are also some striking annual species:—*M. Coulteri*, Gray, is one of the most beautiful flowers of the San Joaquin valley, and is occasionally found as far south as San Fernando; the heads are white or creamy, becoming rosy with age. *M. Californica*, DC., has heads an inch and a half or two inches in diameter, each consisting of between one and two hundred pale yellow ray flowers. There are acres and acres of untilled sandy soil in the south where these regal heads are lifted on their long peduncles above the sunlit carpet of lesser flowers.

The parsley family, Umbelliferœ, contains about thirteen hundred species. Like Compositœ, this order is easily recognized, but the species are very difficult to identify; they do not form a conspicuous part of our native flora, but some very common plants are among them. A little enterprise in observing and collecting would enable older classes to make out a long list of family traits. The clusters are umbellate, often being compound umbels; the individual flowers have inferior ovaries; the calyx limb has usually five minute teeth; there are five petals, five stamens, two styles and an ovary that separates into two akenes; the stem is usually hollow and grooved, and the leaves compound, often finely dissected; carrots, parsley, and fennel are examples; the petioles are sometimes prominent, as in the celery; the roots often store much food, and in some species are edible; frequently the plants are very aromatic, especially the seeds, as in the fennel, dill, coriander and caraway.

Our most notable example of a noxious species is the poison hemlock, *Conium maculatum*, Linn , which blooms in rich, shaded soil in June. Our notable exceptions to the family trait of compound leaves, grow in wet or shaded places. The genus Hydrocotyle, or marsh pennywort, has minute flowers, but very pretty, rounded, scal-

CHAPTER XII

loped leaves, and is a conspicuous feature of the borders of streams and pools in Southern California. *Bowlesia lobata*, R. & P., on shaded hillsides, has exquisite foliage. Our common, early-blooming, field Umbelliferæ belong mainly to the genera Sanicula and Peucedanum. The earliest in Southern California is *P. utriculatum*, Nutt., with compound umbels of yellow flowers much frequented by little insects ; it has finely dissected leaves, and later on conspicuous fruits with red borders. Yellow mats is the name given to some species of Sanicle; there is also a Sanicle with purple flowers. The umbellets of the Sanicle are so compact that they are more properly called heads ; only a few flowers, from five to eight, in each head, are perfect ; the styles of these perfect flowers protrude before the corollas expand, and the stigmas mature much before the stamens stretch out and expose their anthers a few at a time. The little fruits of the Sanicle mature in early summer, and are armed with hooked spines ; the children sometimes call them beggars' ticks. The other Umbelliferæ noted in the Reader, are probably familiar by their common names at least. The botanical names are :—For the wild celery, *Apium graveolens*, Linn.; the caraway, *Carum carui*, Linn,; the carrot, *Daucus carota*, Linn.; the parsnip, *Pastinaca sativa*, Linn.; coriander, *Coriandrum sativum*, Linn.; fennel, *Fœniculum vulgare*, Ger.

The plants of this chapter have many individual traits worthy of attention. As noted in Chapter III, many Compositæ are very successful in adapting themselves to the vicissitudes of a dry climate. The Umbelliferæ generally seem to require more moisture; at any rate several species that are persistent weeds in the Atlantic States do not become troublesome here. Devices for the protection of flowers from moisture or from over radiation of heat, are common in these orders. The flower clusters of some species of Umbelliferæ take a pendent position at night, so that the involucre acts as a tent ; and many Compositæ fold the bracts of the involucre or the corollas of the ray flowers over the rest of the cluster. The methods of seed distribution are varied and efficient. We have seen that a large number of Compositæ have the calyx limb modified to a ring of bristles or hairs, so that the fruits are rendered buoyant ; sometimes the bristles serve to attach the fruits to passing objects. When this calyx limb, or pappus, is easily separated from the fruit, as in the thistle, the fruits that are landed against fences or walls simply detach themselves and fall to the ground, so that we sometimes have hedge rows of Compositæ in such places. Frequently the calyx limb becomes a prehensile organ, as in the genus Bidens ; or again, the involucre, closely investing the akenes, is armed with teeth and hooks, the Xanthium and

SUPPLEMENT

Franseria are examples. In other species, the ripened akenes are held within the persistent involucre, just as seeds are contained in capsules, until the wind is strong enough to scatter them. The fruits of Umbelliferæ have much of their stored food in the form of oil, and so are rather buoyant : besides this, they have frequently thin margins or wings. Other genera have their fruits covered with hooked spines.

But it is when we consider the pollination of the flowers, that the advantages of their close association are apparent. Solitary flowers may develop individuality, and succeed in attracting the most desirable of guests, but nature's preference for the more social method is evidenced by the fact that ninety per cent of all flowering plants have their flowers in clusters. We are told that the two orders discussed in this chapter exceed all others both in the number and variety of their guests. Insects with long tongues rarely frequent the Umbelliferæ, but the honey of the Compositæ, since it is at the base of extremely slender tubes, rising only occasionally so as to be visible in the throat, is reserved for butterflies and bees. The pollen of both orders is abundant, and is gathered by many species of insects.

These orders generally provide for both cross and close pollination. The clusters of many Umbelliferæ are, like the sanicle, in their first stages, practically pistillate, and hence invite cross pollination, but the very viscid stigmas do not wither until the anthers of neighboring flowers begin dehiscence; in most cases, the styles diverge and the stigmas apply themselves to their neighbors' pollen. The fennel, in later stages, sends up its staminate umbellets above those with still receptive stigmas, so that pollen falls on them. A large number of Umbelliferæ have many staminate flowers mingled with the perfect ones. Most Compositæ show a preference for pollination from one head to another. In some cases the plants are actually diœcious; frequently the ray flowers are pistillate and mature some time before the disc flowers of the same head; sometimes all the flowers of a head are in the pistillate stage for a time, and finally all are staminate, but the condition of the sunflower is the most common, that is, there is a shifting ring of flowers in the pistillate stage surrounding those that are shedding pollen. The flowers can usually get pollen from their neighbors in the same head; their guests bring it, or they may stretch out their style branches and reach it, or they are pollinated by the closing of the heads, or perhaps the receptacle is convex and the pollen falls from the younger central flowers upon the stigmas of older ones. Besides this pollination from neighbors, self pollination is not rare; as the style branches separate and coil back, the inner, receptive surface may be brought against the outer, pollen-covered surface, or against the pollen clinging to other parts, to tips of anthers, or to the corolla, or possibly to the long pappus. Of course one will not force all these different conditions upon the attention of the children.

But children can think out for themselves the various purposes served by the involucre, protection from moisture, cold, grazing animals, unbidden guests, and the like. Notice, too, that the involucre makes it impossible for insects to get honey by biting through the corolla tubes.

SUPPLEMENT

CHAPTER XIII.

PLANT FAMILIES. PART I. ENDOGENS OR MONOCOTYLEDONS.

Botanists do not agree upon the classification of **lower plants**, and if there were a settled system it would hardly be taken up to any extent in grammar grades. In a general way, Algæ are considered in Chapter I of the Reader, Fungi in **Chapter IV**, and Archegoniatæ, including Bryophytes and Pteridophytes, in **Chapter VI**. Gymnosperms would have a chapter by themselves if they were generally accessible to school children. Gymnosperms include, besides the Coniferæ, two **orders**, mainly foreign ;—the cycads, resembling tree ferns or palms **and** frequently seen in green houses; **and the order Gnetaceæ**, or joint firs, shrubs or **trees** with **jointed, rush-like stems** ; we have a few species of one shrubby genus on our deserts. Some characteristics of Coniferæ are easily developed, such as their size, general outline, resinous juice, scale-like or needle-like leaves, persistent in all our species, (we have no larches) and their cone-like flower clusters. Representatives of two groups are illustrated in Chapter VIII, and the flowers are described in the Supplement.

Of the cypress group, the Monterey **Cypress has already been** described ; its scale-like leaves and **oblong or roundish cones are** typical ; the group includes, besides the cypress, the cedar, arbor **vitæ and juniper, the juniper differing from** the others in having drupe-like fruits. Introduced specimens of these genera are **common in** cultivation. Our native incense cedar, *Libocedrus decurrens*, Torr., is widely distributed in the mountains ; **under** favorable conditions it attains **a height of one** hundred and **fifty feet, and so is reckoned in** with the California mountain giants that in the Sierras form perhaps the most remarkable forest belt in the world. The incense cedar in its prime is a beautiful, spicy, symmetrical tree, its downward curving branches being described as great, " ferny plumes." Our native junipers, with thick, sturdy trunk and limbs, adapt themselves to high, rocky and arid regions.

CHAPTER XIII

The great Coniferæ of our forests are so famous, our mountains are so accessible, and it is so desirable to awaken or foster a love for them, that it seems to me quite legitimate to teach children some of their most notable trees from books and pictures, and from any fragmentary specimens, such as cones and the like, that may be available. It would be a dull teacher indeed who would fail to arouse enthusiasm with such a book in hand as John Muir's "Mountains of California," with its chapter on forests. The "big tree," *Sequoia gigantea*, Decaisne, is, perhaps, of the greatest interest. Handsome young trees of this species are often seen in cultivation, but the native forests exist only in the Sierras at altitudes of from five to eight thousand feet. The leaves are scale-shaped, but have long, pointed tips. The cones are absurdly small, only about two inches long, and have a "quilted appearance," as Miss Eastwood observes. The bark of the older trees is fibrous and very thick ; fragments of it are sold at curio stores. The tallest tree reported is three hundred and twenty-five feet ; Muir mentions one with a diameter of thirty-five feet and eight inches inside the bark four feet from the ground. Trees are known that exceed these in height or in diameter alone, a Eucalyptus over four hundred and fifty feet in height, and a chestnut over sixty feet in diameter being examples ; but, taken as a whole, our tree is probably without a peer. The age of our Sequoias is a matter of dispute, but some of those now standing are believed to have been flourishing trees before the Christian era. These mere figures will, of course, mean little to children unless vivified by descriptions and by comparison with known objects. The pamphlets advertising routes of travel to these sections supply some striking comparisons. The following bit of description is from Muir : " The young trees have slender, simple branches down to the ground, put on with strict regularity. By the time the sapling is five or six hundred years old, the spiry, juvenile habit merges into the firm, rounded, dome-form of middle age, which in turn takes on the eccentric picturesqueness of old age." That millions of seeds are ripened in a single year is literally true, and although the squirrels may have ninety-nine out of every hundred, there are still more than enough to keep our forests well supplied with new trees, if only the devastation of shepherds and lumber men can be checked.

Great forests of Redwood, *Sequoia sempervirens*, Endl., occupy our coast ranges from Oregon to San Luis Obispo. These trees, too, are giants, being from two to three hundred feet high and from eight to twelve feet in diameter. The children are pretty sure to know something of their value. Pictures and descriptions of lumbering

SUPPLEMENT

among the redwoods can be found in "Picturesque California," page 453.

The Douglas spruce, *Pseudotsuga Douglasii*, popularly called the Oregon Pine, grows in dense forests in Oregon and Washington, where it is frequently three hundred feet high. It is also widely distributed in California, its variety *macrocarpa* extending to the very southern counties. In Los Angeles County it is found at lower altitudes than the other Coniferæ. It is not a true pine; its leaves are linear, not needle-like; they are two-ranked and are petioled. The most characteristic feature of the tree is its cones, which have fringe-like bracts overlapping the scales. Of the true pines, the most majestic is probably *Pinus Lambertiana*, Torr., popularly known as the sugar pine because of the sugar that exudes from its heartwood when wounded. This tree, too, ranges from one hundred and fifty to three hundred feet in height and from ten to twenty in diameter. It can be readily recognized from its cones; they are from fifteen to eighteen inches in length and, when open, from four to five inches in diameter. Its needles are in fascicles of five, but are only about three inches long; the bark is broken by fissures into small scales or plates. The bark of the yellow pine, *P. ponderosa*, Dougl., on the other hand, is in massive plates; its needles, eight inches long, are in groups of three, but its cones are only about four inches long. This pine rivals the other in height and has a much wider range, being by far the more common in the mountains of Southern California. Both of these pines are prominent in the Sierra forest belt, and are vividly characterized by Muir. Also of this illustrious company of mountain kings, are the silver firs, *Abies concolor*, Lindl., and *A. magnifica*, Murr., slender, beautifully symmetrical, silvery trees, hardly less than the others in height, "the younger trees dressed with such loving care that not a leaf seems wanting."

From our list of Coniferæ, the nut-pines should not be omitted. *P. Sabiniana*, Dougl., grows on the hot foot-hills of the western slopes of the mountains; it has sparse foliage but very large cones. *P. monophylla*, T. & F., is the nut-pine of the eastern slopes of the Sierras and the arid regions beyond. The cones of this species are very small but are full of nuts; Muir believes that in fruitful years the crop of these nuts exceeds the wheat crop of California; his account of the gathering of the nuts by the Indians would be sure to appeal to school children, to whom the nuts or piñons are familiar objects. The coast, or Monterey pine, *P. insignis*, Dougl., has been already referred to as our most common cultivated pine. Its leaves are in threes and are from four to six inches long; the cones are pointed and curve inward; they remain on the tree several years without opening.

In classifying Angiosperms, the system adopted in Campbell's text book, that of the German botanist, Eichler, is followed. In studying seedlings, Chapter II, the distinction between monocotyledons and dicotyledons was noted. In their earliest stages, both exogenous and endogenous stems consist of pith in which woody fibers are scattered, but the woody bundles of the exogen soon form a ring surrounded, as noted in Chapter VIII, Supplement, by a cambium layer that adds each year new rings of wood and bark. In endogenous stems, the woody bundles remain scattered throughout the pith, and, while near the circumference they are more numerous, rendering the tissue firm and strong, there is no cylinder of wood, no true cambium, and hence no rings of annual growth. Some endogenous stems do not increase at all in diameter after the outside tissues are fully formed, a few increase slowly by the interposition of woody fibres, but by far the greater number of endogens have no perennial stems above ground Short underground perennial stems, that is, rootstocks, bulbs, corms and the like, are very common in this group. Other common characteristics, as the children will discover for themselves, are entire, parallel-veined, and often, long, vertical leaves; and flowers whose parts are in threes.

Of the group Liliflorœ, the order Liliaceæ alone has two thousand species, many of these common in cultivation, so that material for the study of family traits is not difficult to obtain. The lily-of-the-valley, tulips, crown imperial and some other favorite garden plants in more humid climates, are rather rare in California; but several of the lily-like plants enumerated in the Reader are common. The smilax, or more correctly the Mersiphyllum, blooms very freely early in the year; its little star-like, white lilies have a delicious fragrance, and attract guests in throngs. The foliage of this plant is unique; the true leaves are reduced to mere scales in whose axils grow the flowers and also the apparent leaves, which are really modified branches. Besides the native lilies mentioned in Chapter VII and its Supplement, and the Mariposa, Chapter IX, some others may have been encountered before the children reach this chapter. There are several very handsome species of native Brodiæas, some of them now under cultivation; *B. minor*, Wats., with umbels of large violet flowers is rather common in Southern California; *B. laxa*, Wats., sometimes called Ithuriel's spear, has still larger umbels; it is rare in the south, but fairly common in rich, shaded soil further north; some of the species of Brodiæa are white, yellow, and even scarlet,

SUPPLEMENT

but they usually bloom late, and only on the more northern hills and **mountains.** The golden-star lily is *Bloomeria aurea*, Kell.; its tall, **showy** umbels deck many **a** dry hillside **after the greater part of the spring** vegetation, including **its own leaves, has withered.** The flowers, which seem so **openly hospitable, really exclude all but bees and butterflies** from their **honey by means of peculiar appendages at the base of the stamens.** The soap-root will **be taken up in Chapter** XV, also some **of** the genus Lilium that bloom in **summer.**

The Yucca **of the** illustration is *Y. Whipplei*, Torr., which, with **its** variety *graminifolia*, is common throughout the southern foot-hills. After **some years** of preparation, the rootstock **with** its clusters of **bayonet-like leaves, sends up** a flower stem which, in **a few** weeks, **attains a height of from ten to fifteen** feet and bears literally **thousands of** flowers, **forming** one of the most beautiful flower panicles **in nature.** The tree Yucca, or "Yucca palm," of the Mohave Desert is *Y. arborescens*, Torr., a remarkable plant in its details, as well as in general appearance. From its base countless roots, slender as whip cords, radiate in every direction, extending long distances but lying near the surface. Its trunk and branches are for a long time clothed with savage, reflexed leaves, but when these finally fall, the stem **is** seen to have acquired bark of considerable thickness, **and a section of it** shows concentric rings **characteristic of exogenous** rather than endogenous stems. **The growth of this Yucca stem and** of another phenomenal **endogen, the dragon-tree of W. Africa, which has** been known to **attain seventy feet in** height and sixteen feet **in** diameter, has received much attention, but the subject is beyond the scope of this **work.** *Y. Mohavensis*, Sargent, with fleshy, purplish fruits, **has a wide range in** California and in the South West **generally, and is extremely** variable in form.

The pollination of the Yuccas has long been a subject of close **observation and study by** both botanists and entomologists. **The** story as told in the Reader applies to any of the species. The **three** common California species have been individually investigated; they are pollinated by three different, but nearly related species of moths; their methods of procedure vary to some extent, but the leading facts are the same in all cases. The moths that pollinate *Y. Whipplei* and its variety may be frequently seen at work before **dark.** Most species of Yucca mature their stigmas before the **anthers, and there** is always a stigmatic cavity (very slender indeed in Whipplei); **the** moths always seek newly opened flowers for **depositing** their eggs, so **the** pollen they bring is nearly always from **other** flowers. **The** moth **does not seem** to require food during her **brief existence as a** moth;

at any rate, she has never been detected eating. She spends considerable time at the stigma forcing the pollen down the stigmatic cavity; usually she deposits an egg in each of the six seed divisions of the capsule, and repairs each time to the stigma to deposit pollen. The adaptation of the moth's antennæ and ovipositor to the conditions of pollination is perhaps the most wonderful part of the story. Some species of Yucca secrete a little honey, but it seems to play no part in pollination. There may be occasionally accidental pollination without the agency of the moth, but ordinarily, if the Yucca produces fruit, the moth is found on the flowers, and its larvæ in the fruits. Summaries of the studies on the Yucca were published in the annual report of the Missouri Botanical Garden issued May 1892 and March 1893. A brief popular account can be found in the Popular Science Monthly, Vol. 41.

Our Yuccas are utilized in various ways by the Indians. Not only the fleshy fruits, but the seeds of the dry fruits, and even young flower shoots are eaten; the woody fibre is used for cordage, horse blankets, hats, baskets, &c., and macerated sections of the stems serve as a substitute for soap; in these latter days sections of pith from the larger stems are put to various uses by tradesmen, artists, surgeons, scientists and manufacturers.

The century plant, *Agave Americana*, is indigenous to Mexico, and the natives have found even more uses for it than our Indians have discovered for the Yucca. Subterranean stems and leaf fibres serve similar purposes in both cases, but the Mexicans cut off the central bud of the Agave, just as it is about to develop into the flower cluster, and collect the abundant sap that the plant has so long been preparing for its supreme effort; by fermentation or distillation, they make of this juice their national drinks, pulque and mescal. California children frequently have opportunity to watch the marvellously rapid growth and development of the Agave flower stalk, and to observe the throngs of winged creatures that drink honey from the overflowing flowers.

Most introduced Liliforæ have provisions for cross pollination so obvious that children can find them out for themselves. Of course the pollinating agent of their native country may not exist here. The rushes, Juncaceæ, like the lilies, have flowers with parts in threes, but they are inconspicuous and are adapted to wind pollination. The Spanish moss, which so beautifully drapes the trees of our South Eastern States, and the pine-apple of commerce, also have flowers with parts in threes, and are usually grouped here; the former has the appearance of a lichen, but is really a leafless epiphyte living on moist-

SUPPLEMENT

ure and organic matter dissolved therein, and producing true flowers and seeds; what we eat of **the** latter **plant** is the consolidated flower cluster, that is, the individual fruits **and** their stems grown juicy. The Tradescantia, or **wandering-Jew,** of cultivation, **is** usually assigned to another but **a nearly related group.**

The next group **considered in** the Reader is known **as** Spadiciflorae, from the spadix-like inflorescence. Duckweeds, Lemnaceæ, are minute floating plants, common **the** world over on quiet pools; **each plant is a** small **green disc with,** usually, one descending root; **they are rarely found in flower and then the** flower **is** reduced to a single **stamen or pistil**; **ordinarily they** reproduce rapidly by division. **There** are **various other floating plants of** the order Naiadaceæ, which are commonly **assigned here**; **they are known as** pondweeds, ditch grass, &c., **and they choke up our reservoirs and irrigating** ditches. The calla, *Richardia Africana*, **is a type of a group of about** one thousand **species,** nine-tenths of them tropical. **They are** known as Aroids: "**Jack in** the pulpit," sweet flag and **skunk** cabbage, common in our Eastern States, belong here; so does our cultivated black lily with its livid colors and fœtid odor that attract carrion-eating insects. The tropical aroids sometimes attain most luxuriant growth; one from the Island of Sumatra produced in Kew Gardens, **London, a** spathe six feet long and about three **in circumference, with leaves and rootstock in** the same proportion. There are climbing **aroids that clamber over the tops of forests and send down aerial roots, which fasten themselves in the soil.**

There are about **a thousand known** species of palms also. Two **genera are native to California, one** species, *Washingtonia filifera*, Wend., **being common in cultivation;** a tree of **this species, in Los Angeles,** known to **be fifty years old,** is sixty feet high. **Several** foreign **palms, too, thrive well out of doors in** our climate, and since palms **are so** typical of the vegetation **of the** tropics, and have there so great **economic value,** it is desirable to emphasize the group in connection **with Geography work. As can** be seen from our own palms, the inflorescence **is** usually crowded, consisting of great numbers of small flowers, sometimes **perfect, sometimes** staminate or pistillate only. The fruits may be berries, drupes **or nuts. Palms growing in a** very **hot** and humid **climate must have** enormous **leaf surface for sufficient** transpiration. The following figures are taken mainly from Kerner: A Brazilian palm has pinnate leaves seventy **feet long and twenty-five feet** broad; the Talipot palm of Ceylon **has palmate leaves eighteen by** twenty-five feet, **and other** parts in **proportion; in** the course of **forty years, its** trunk attains a height **of about seventy feet; it** then

CHAPTER XIII

sends out from the top an inflorescence forty feet high and thirty-five in diameter; after maturing the fruit from this cluster, the Talipot, like the century plant, dies. But there are other palms that mature great abundance of fruit every year. The cocoa nut palm, now cultivated throughout the tropics, yields from one hundred to one hundred and fifty nuts a year for perhaps forty years. The fruit of the date palm constitutes the main food of the inhabitants of the Arabian and African deserts. Sometimes the main food supply is stored in the trunk of the palm; in the East Indies the pith of a sago palm fifteen years old has been known to supply eight hundred pounds of sago. The fruits of a West African palm supply the palm oil of commerce; the nuts of others furnish a substitute for ivory; several kinds of palms have their young leaves coated with a wax of commercial value, and so on. The multitudinous uses the natives find for cocoa palms can be looked up, as suggested in the Reader, and are sure to interest children. Palms have been known to attain a height of one hundred and eighty feet, but the greatest measurement attained is by some of the climbing palms; Kerner says that a length of six hundred feet has been recorded.

There are thirty-five hundred species of grasses, and some of the species are so very widely cultivated that this number fails to indicate the importance of the family. Family traits are easily seen. The leaves are linear and have sheathing bases; the stems are jointed, and in most species are hollow between the joints; the flowers, when perfect, have usually three stamens and two very feathery stigmas, but many grasses besides the corn have unisexual flowers. The flowers rarely show traces of a perianth, and the usual arrangement of bracts surrounding them can be easily seen in the wild oats, or in the staminate flowers of the corn. An inner bract very closely invests the flower and an outer bract nearly envelopes this flower and the inner bract; finally two or three flowers, or florets, with their bracts are together enclosed in another pair of bracts called the glumes; these groups, called spikelets, are clustered in various ways; they form a panicle in oats, a spike in barley, and so on. The root system of the corn seedling, Chapter II, is typical of annual grasses. In tropical countries some grasses, like the bamboo, have perennial stems that attain the height of trees; the bamboo is sometimes seventy-five feet high. In temperate climates the vertical stems of most perennial grasses, unless artificially treated, die down annually, only the long horizontal stems, above or below ground, being perennial; Bermuda grass and the Kentucky blue grass of our lawns are examples, also the perennial grasses that persist in our moist alkali lands.

SUPPLEMENT

The strength and flexibility of the stems of grasses have been already noted. The leaves are usually well adapted to their environment, the species in humid climates being able to shed water because of lines of wax, while many genera of drier regions have the power of folding and unfolding their leaves to meet the varying conditions of the atmosphere. Kerner devotes several beautifully illustrated pages to this latter habit. The devices for the pollination of grasses are very interesting, but they require close observation. In general, the pair of feathery stigmas protrude from their wrappings for some time before the anthers are allowed to escape; the bracts separate to allow the exsertion of the anthers only under favorable conditions as to temperature and moisture; as soon as the anthers are exposed, the filaments lengthen almost as if by magic, attaining their full length in a few minutes; the pollen is shaken from the suspended anthers in the course of perhaps half an hour, though the empty cells may remain for a day or so. The fruits of grasses, or rather the bracts that invest them, are frequently provided with barbed bristles, or awns, that aid in seed distribution. The awns of several genera besides the oats are hygrometric, the motion being spiral in the genus Stipa. These awns, of course, serve the same purpose as the styles of the filaree fruits.

Identification of grasses is exceedingly difficult, but we have an illustrated flora, the "Grasses of the South West." California has some nutritious native grasses known as bunch grasses, but generally our native grasses are not abundant and are yielding place to foreign species, the most notable of which will be described in Chapter XVI. No plant family can be compared with the grasses in economic value. Rice, now universally cultivated in the tropics, feeds millions of the human race, and wheat comes next. Wheat is supposed to have originated in Asia, but it has been cultivated in Europe from prehistoric times. Barley and rye are probably of European origin; Indian corn is American. Oats are of special value in countries with a brief growing season. The cultivation of the sugar cane is still a leading industry in warm countries; sorghum is nearly related to sugar cane, and so is broom-corn The uses of the bamboo to the natives of tropical Asia are well known.

The Canna-banana group is tropical, but should be studied because of its geographical importance. Sections of the apparent stems of the banana, its flower clusters and young fruits are frequently accessible. The apparent stem consists mainly of sheathing petioles; the dark red bracts constitute the showy part of the flower cluster; individual flowers have inferior, three-celled ovaries, six-parted irregular perianths, and five stamens with some traces of pollen. Like many other highly cultivated plants, bananas have lost the power of producing fertile seeds. The banana and the nearly related plantain are the staple food plants in many parts of the tropics; their productiveness in suitable soil is almost incredible; it is said that an acre can produce sixty-six tons of fruit, a fact that helps to explain the indolence of the natives.

The Orchid family is next to the Compositæ in number of species, the estimate being eight thousand. The plants are most interesting in their habits; many of the tropical species are epiphytes in the tops of forest trees, others are saprophytes. The flowers are, of all known species, the most highly specialized with reference to insect visits. But California is very poor in orchids, so that actual observation work with them would be rarely feasible, and they are not of sufficient economical importance to warrant study from books on that pretext; so little space is assigned them here. Orchids have an inferior, one-celled ovary and a six-parted perianth, one part of the perianth, called the lip, being very conspicuous. Most species have but one fertile anther; the lady-slipper is an exception, having two. The anthers are usually united with the pistil. The lady-slipper, Cypripedium, a genus not rare in the north, admits guests only where they must strike the stigma; once within the banquet hall, that is, the lower lip, the guest's only exit is via the anthers, and these exits are so narrow that he must smear his shoulders with pollen. The genus Habenaria is the one that compels its guests to carry pollen masses on their eyes; there are several California species of Habenaria bearing spikes of small white or greenish white flowers, but they are not abundant and would rarely be noticed; those I have observed have been fully pollinated. Illustrations of a British Habenaria and a full description of its pollination are given in most botanical text books. The orchid referred to in the Reader as hurling its pollen masses, is *Catasetum saccatum*, fully described by Darwin, also by Kerner. Orchids have the smallest of all seeds; so minute are they that, like spores, they may remain suspended in the atmosphere, and so are widely distributed.

SUPPLEMENT

CHAPTER XIV.

PLANT FAMILIES. PART II. EXOGENS OR DICOTYLEDONS.

The characteristics of Dicotyledons have already been indicated in Chapters II and XIII, and the structure of perennial exogenous stems has been given in the Supplement to Chapter VIII. There is little observation work with new plants suggested in this Chapter, its aim being mainly to review systematically plants already studied, and to aid in Geography work.

The members of group I come under the Apetalæ, if this division is made. Many of the native trees of the group have been already considered in Chapter VIII and its Supplement. The willow belongs to the Northern Hemisphere, and extends well up into arctic regions. The two species of Populus that grow along our streams, *P. trichocarpa*, Torr., and *P. Fremonti*, Wats., are indiscriminately called cottonwood, or poplar; the mountain species, *P. tremuloides*, is known as aspen. Our fine park or field live oak, the Spanish encino, which is common near the coast, is *Quercus agrifolia*, Nee.; near Los Angeles this is interspersed with *Q. Engelmanni*, Greene. *Q. chrysolepis*, Liebm., of the mountains sometimes attains immense size; the noble white oak, or Roble, *Q. Lobata*, Nee., with its deciduous lobed leaves, is found in fertile valleys, especially northward; *Q. densiflora*, H. & A., the chestnut oak, or tan bark oak, has bark so valuable for tanning leather that the species seems doomed to extinction. Oaks throughout the North Temperate zone are prized as ornamental and timber trees; several species are valued for tanning; the cork oak of Mediterranean regions is sometimes seen in cultivation in California. While the walnut of the old world can be very profitably cultivated in parts of California, the native walnut does not compare in size with that of our Eastern States. There are several other members of this group much prized for their nuts and wood which are common in the Eastern States but do not thrive here; the chest-

CHAPTER XIV

nut, hickory and beech are examples. The birch, whose bark is generally known, is very rare in California, and we have to do without the elms, that are such a feature of Eastern and European landscapes. Our alders are specially fine, being trees rather than shrubs.

The fig, or Ficus, family is a notable one in the tropics. Many species have remarkable roots. The India rubber, *Ficus elastica*, sometimes attains sufficient size in our climate to show its buttress-like roots. Other species of Ficus have climbing, lattice-like roots, which the natives sometimes convert into living bridges. Still others send down from their branches many aerial roots which fasten themselves in the soil and become sturdy columns, resembling trunks and serving the same purpose. The banyan trees of India are of this sort, and are like massive roofs supported by hundreds of columns. The sacred tree of the Hindoos is *Ficus religiosa;* one tree is said to have sheltered an army of five thousand men, and in Ceylon there is a tree of this species, beneath which a village of one hundred huts is built. Some species of Ficus have juice that is poisonous as well as gummy or milky; the upas tree of Java, the subject of many fabulous stories, does furnish a deadly poison which the natives use on arrow heads.

The hop, hemp and mulberry plants are all akin to the fig. The bitter principle of the hop is developed in the pistillate flower clusters. The hemp, besides furnishing fibre for cordage, also yields various narcotics, hashish and the like. The leaves of several species of mulberry are food for silkworms. The berries are the flower clusters, perianths, stems and all having grown juicy. The breadfruit tree of the tropics is nearly related to the mulberry; its aggregate fruits are baked just before maturity. The nettle family is a large one, and is widely distributed; a few members, the ramie for instance, are valuable fibre plants. The tree nettle is from fifty to one hundred and fifty feet high. There is a curious Australian tree family, Casuarinaceæ, belonging to this group; one species is cultivated to some extent in California; it is leafless, but has slender, drooping, jointed branches resembling those of the Equisetum.

The yerba mansa is *Anemopsis (Houttuynia) Californica*, B. & H. The involucre is petaloid and conspicuous, and each little, fleshy flower is subtended by a minute white bract; the plant has a pungent odor. There are several unique plant families that are sometimes assigned to this group and sometimes left unclassified; the mistletoe family Loranthaceæ; the parasitic Rafflesiaceæ, and the family Aristolochiaceæ; the last is mainly tropical, but two species, the Dutchman's pipe, *Aristolochia Californica*, Torr., and the wild ginger,

SUPPLEMENT

Asarum caudatum, Lindl., grow somewhat sparingly in our mountains. Some small species of the Rafflesiaceæ are found in our arid regions, but in the tropics the flowers of this group attain monstrous size, as was noted in Chapter IV. Mistletoe is very common on the oaks, the sycamores and junipers, and to some extent on other trees, in our cañons and mountains. It draws nourishment from the host plant by means of little modified rootlets, called sinkers, which reach into the newly forming wood cells. The flowers are inconspicuous; the berries are familiar objects at Christmas time; the seeds are, of course, distributed by birds.

Six of the seven families that constitute the next group are well represented in California; four of them are entirely apetalous, but the highest two families have some apetalous genera and other genera with complete flowers. Our representatives of the order Nyctaginaceæ have showy calyxes. The four-o'clock, *Mirabilis Californica*, is described in the Supplement to Chapter VII; the sand verbenas, Abronias, are common on our beaches and deserts, and are interesting types of the plants of arid regions. The family Amarantaceæ is too well represented by the tumbleweeds; Chenopodiaceæ, by the goosefoot, pig-weed and the like; Polygonaceæ by dock, canaigre, knotweed and Russian thistle, all of which weeds will be considered in Chapter XVI. The beet, spinach and the rhubarb, or pie plant, of our gardens belong here. The sugar beet is only a variety of the common beet, *Beta vulgaris*. The Eriogonums, described in Chapter XV, belong to the family Polygonaceæ, so does the genus Chorizanthe; species of the latter genus grow very profusely in sandy soil; the clustered leaves appear early in the year, but the prickly white or pink flower clusters do not come until early summer, when the low plants are much branched, and are dry and brittle. The pink family, Caryophyllaceæ, is a large one, and is especially well represented in northern or mountain regions; the weedy members, chickweed, sand-spurry and catchfly, are noted in the Supplement to Chapters VII and XVI. Our handsomest native species, *Silene laciniata*, is described in Chapter XV. The family Portulacaceæ has comparatively few species; the purslane (Chapter XVI), miner's lettuce (Chapter VII, Supplement) and Calandrinia (Chapter IX, Supplement) are our most common representatives.

The families of the third group of Choripetalæ, are included in the Polypetalæ of the older books, all but a few genera having both calyx and corolla. The lower families have all their parts free and distinct, the higher have compound pistils. The family Ranunculaceæ, usually placed first or lowest, has no union of parts; the peony and buttercup are considered in Chapter VII; the larkspur and columbine in Chapter XV. Clematis and meadow rue, common in cañons, have no corolla, and some species have unisexual flowers. The calyx of the clematis is showy, being white in our native species, purple in some introduced species; the fruits with their long plumelike appendages (the styles) are conspicuous in autumn. The leaves of the meadow rue are tri-compound, and are somewhat like the maiden hair fern; the flowers are greenish, and are adapted to wind pollination. The barberry, while not rare in California mountains, is not commonly known. The laurel or bay, *Umbellularia Californica*, Nutt., is a handsome tree, common not far from water in mountain cañons; it belongs to the laurel family and is nearly related to the bay or laurel of Southern Europe, whose leaves, formerly used to crown heroes, are now put to more prosaic uses, such as flavoring gravies, puddings, dried figs and the like. Nutmeg is the seed of an East Indian tree; mace is the envelope of this seed; cinnamon is the bark of a tree of Ceylon; cassia bark and buds are also from Ceylon; camphor is the gum of a tree of China and Japan; sassafras is a common shrub in our Eastern States.

The magnolia, and several other ornamental and timber trees, belong to this group. There are several aquatic families in this group, but they have few representatives in California. The lotus, Nelumbium, is common in the ponds of our parks; its nut-like fruits are imbedded in a top-shaped receptacle. The *Victoria regia* of the Amazon valley is noted for the size of its leaves and flowers; the leaves frequently exceed six feet in diameter, and the flowers are sometimes a foot across.

The mustard and poppy families have been pretty fully treated in Chapter IX and its Supplement; the opium poppy is extensively cultivated in India, the crude opium being the juice that exudes from slits cut in the green capsules. The caper family, Capparidaceæ, very nearly related to the Cruciferæ, has a few representatives in California; one of them, *Isomeris arborea*, Nutt., is a low, strongly scented shrub, common near the sea; it has yellow flowers and inflated pods on stipes; the capers of commerce are the unopened flower buds of a European species. The Dicentras, "bleeding hearts," and golden ear-drops, belong to another nearly related family, but they are rather rare in California.

Violets and mignonette belong to this group; also some very interesting insectivorous plants, the sundew, the Venus fly-trap and pitcher plants. The sundew, Drosera, and a very remarkable pitcher plant, *Darlingtonia Californica*, are found in remote moun-

SUPPLEMENT

tain districts, but would rarely be available for class work. The plants are illustrated, and their methods described in nearly all textbooks of Botany; the Darlingtonia is treated at length in Kerner's work, Watson's "California Flora" and Miss Parsons' "Wild Flowers of California." It is not easy to condense the stories, so they are omitted here. It is the leaves, not the flowers, that capture the insects, and in most cases there is actual digestion of the prey.

The highest families of this group are the Malvaceæ, referred to in the Supplement to Chapters VIII and IX; the Linden family, which includes the linden, or lime tree, of northern Europe, the basswood of our Eastern States and the chocolate tree of tropical America; and another family which includes the tea tree and the Camellia. Chocolate is made from roasted seeds, tea is, of course, dried leaves. The fruits of the cotton plant, showing the seeds clothed with cotton are easily obtained. The products of the cotton seed, i. e., oil, cottolene, etc., should be noted.

The relationship of the different families of the next group of Choripetalæ is not sufficiently obvious to warrant much attention. The geranium family is rather fully considered in Chapter X, and our common species of Rhus, Rhamnus and Ceanothus in the Supplement to Chapter VIII. The poison sumach and poison ivy of our Eastern States are species of Rhus. Some of the best known ornamental and timber trees of other climates belong here; the numerous maples, box-elder, horse-chestnut, buckeye, the true holly, and many trees of the tropics and Southern Hemispheres, the small tree that furnishes maté, or Paraguay tea, being the most important economically. The citrus trees have been introduced mainly from India; our cultivated grapes are mostly from Europe, but the most common grapes of our Eastern States have been cultivated from native species. Linseed oil, as well as linen, is a product of flax.

No characteristic that is universally true of Euphorbiaceæ can be affirmed, so little attention should be given to the kinship of our scattered species. They are all considered elsewhere in the Supplement, and Euphorbias of special economic importance are sufficiently indicated in the Reader.

The highest group of Choripetalæ is characterized in the Reader, and many of its families are so well represented with us that they are treated specially in other chapters: Umbelliferæ in Chapter XII, Onagraceæ in Chapters IX and XV, Cactaceæ and the nearly related succulent orders in Chapter XV, Saxifragaceæ in Supplement to Chapters V, VIII and IX, Rosaceæ in Supplement to Chapters VIII and XV, and Leguminosæ in Chapter X. There are two families

CHAPTER XIV

nearly related to the Umbelliferæ, but they are not well represented in California. The dogwood family, or Cornaceæ, has a few species along our streams; the flowers are white, and grow in large cymose clusters, the northern species being very handsome. The only familiar representative of the other family is the English Ivy. The edible fruits of the rose family are pretty thoroughly discussed in the Supplement to Chapter VIII. The rose hip is a fleshy, urn-shaped receptacle in which many hard akenes are borne, i. e., it differs from the strawberry in having a concave instead of a convex receptacle. The fruit of the California holly, also a member of the rose family, is perhaps strictly a pome, but it is apparently a berry. The service berry of the mountains, Amelanchier, has also a pome fruit. From the children's list of Leguminosæ of economic value, the peanut should not be omitted; the nut is, of course, the legume; the plant naturally buries its fruit as a protective measure, and this habit is preserved in cultivation. Among timber trees are a sort of rose wood, many kinds of Acacia and the locust. The indigo plant of India is a member of this family; the dye is obtained from the macerated leaves by a process that always interests children.

Plants whose flowers have their petals united are now called Sympetalæ, the old name was Gamopetalæ or Monopetalæ. The heath group is a large one, including over two thousand species, some of which adapt themselves well to rocky mountain slopes, others to sandy plains, and still others to swamps and moorlands. The manzanita and madrone of our mountains are referred to in Chapter VIII; the snow plant in Chapter XV; several other saprophytic members of this family are common in our high pine forests. Our beautiful azalea, *Rhododendron occidentalis*, Gray, is found along mountain streams in summer; another Rhododendron, known as the rose bay, is found in northern woods. We have a native huckleberry, but it is not common. The introduced heaths of our gardens are mainly from South Africa. The primrose group is also sparingly represented with us. The Plumbago is common in gardens, and there are a few beach species. The pimpernel, *Anagallis arvensis*, Linn., is common in the vicinity of cultivation, and blooms the year round; its little salmon-colored flowers are very inhospitable, and must usually pollinate themselves.

The five families of sympetalous flowers that make up the group Tubifloræ, have been fully considered already. The largest of these families, the Solanaceæ, is the one of greatest economic value, including as it does, the potato, tomato, egg plant, tobacco and some other plants that yield narcotics. The sweet potato belongs to the

SUPPLEMENT

family Convolvulaceæ, and a few of the Borraginaceæ furnish dyes. Polemoniaceæ and Hydrophyllaceæ are small families of no small economic value. On the other hand, some of our most attractive introduced plants, as well as native ones, belong to this and the following group. The Petunia and showy species of **Solanum** and Datura are **Solanaceæ**; the phlox is nearly related to the Gilias, and there are some exquisite forget-me-nots and borages. The bilabiate group comprises, besides the **families** Labiatæ and Scrophulariaceæ treated in Chapters XI and XV, the **Verbena**, Acanthus and Bignonia families and a few others; some of the **cultivated plants** belonging **to this group are** the verbenas, lemon verbena, **catalpa**, Gloxinia, **foxglove and the** beautiful Japanese tree, *Paulownia imperialis.* The economic value of this group is very small. Some members of the mint family, rosemary, lavender, thyme, hoarhound, and the like, are put to minor uses.

Botanists do not agree in the placing of the members of the next two groups. The ash of Europe and our Eastern **States is** nearly related to the olive. The gentian family is a large one, **but it is** confined mainly to cold climates or Alpine regions; we have a few beautiful species in our own mountains. One member of the gentian family, *Erythræa venusta*, an annual **having very bright pink flowers with** twisted **anthers, is common in sandy soils in our valleys in May and** June. The gentians are supposed **to possess some medicinal value**; **a nearly related** tropical **family furnishes strychnine and other** poisons, some of **them** used by the natives **for** tipping their arrows. The milkweed family, though sparingly represented in California, is **a** very large **one.** Two rather common cultivated plants belong here, the wax plant, *Hoya carnosa*, and **a** cactiform **plant, Stepilia, from South** Africa **; they can be** readily recognized **by the peculiar** anthers. The dogbane, oleander and periwinkle are nearly related to the milkweed family. **We have** no common, conspicuous Campanulas excepting the introduced Canterbury bell; the brilliant cardinal-flower, *Lobelia cardinalis*, is rare. The rank of the family Cucurbitaceæ is by no means settled; our most common representative, the chilicothe of Chapter V, has unisexual flowers that are apparently apetalous, but the perianth may be considered as **a** blended **calyx and corolla**. The origin of melons, pumpkins **and** the like **is also** a matter of dispute.

Compositæ, the most important family of **the** group Aggregatæ, **is the subject of several pages of** the Supplement to Chapter XII. It **is of surprisingly little economic** value, but **some of the allied families are more useful. The** family Dipsaceæ furnishes the fuller's teasel used in the manufacture of woolens; the teasel often escapes from cultivation and becomes a troublesome weed. The family Caprifoliaceæ contains the elder as well as the honeysuckle and snowberry, all noted in the Supplement to Chapter VIII. The family Rubiaceæ has but one California genus, Galium, consisting of low, weak, square-stemmed plants with leaves in whorls, and small greenish flowers. This genus does not well illustrate the kinship of this family to Compositæ; the relationship is more apparent in some other Rubiaceæ, the cultivated button-bush for instance. Two other genera, the Cinchona of South America and the Coffea of Abyssinia, because of the alkaloids they contain, have become of great commercial importance, and are now widely cultivated in tropical countries. Quinine is obtained from the bark of the Cinchona. The coffee is a small evergreen, producing berries that contain two seeds each. All the coffee of commerce is from the same species, but the seeds are of different quality in different climates. It has been cultivated in Arabia for five centuries, and in Java for two.

SUPPLEMENT

CHAPTER XV.

SOME SUMMER FLOWERS.

From May until the autumn rains, the prevailing brownness of the California landscape is likely to mislead one into laying aside all **study of living plants, and so** into missing some of nature's most interesting pages. The soap-root, *Chlorogalum pomeridianum*, Kunth., **is one of the** earliest plants to respond to the autumn rains, as noted **in** Chapter V. This plant is very widely distributed, and it is easily studied, since its flowers expand nearly on time when the clusters are kept in water indoors.

The larkspur, Delphinium, belongs **to** the family Ranunculaceæ. The species are many and **are** difficult **to identify, especially the blue** ones. The **one** in the **illustration is** *D. Parryi*, Gray, **common on hillsides in the vicinity of Los Angeles;** there are also earlier species, **and in the mountains in summer** there **are blue larkspurs with exceedingly** tall, handsome clusters. The scarlet larkspur of the south, *D. cardinalis,* **Hook.,** is larger and handsomer than the northern, *D. nudicaule;* **it is** a magnificent plant, takes kindly to cultivation, **and deserves** more recognition than it has received. **Only larger** bees, the carpenter and the bumble-bee, seem **able** to get **the honey from** the blue larkspurs that I have observed. It will be noted that larkspurs use exactly the same device as the cultivated nasturtium in holding their anthers before the entrance to the honey during their dehiscence, afterwards replacing them with matured stigmas. Individual flowers last from a week to ten days, and the clusters are of very long duration.

The milkweed of the illustration is *Asclepias eriocarpa*, Benth , very common in June in sandy wastes in the south. *A. Mexicana,* Cav., a more slender plant with smooth leaves and smaller flowers, is common throughout the state; several other milkweeds are likely to **be** encountered, but their story of pollination is substantially the **same. Within the honey sacs are horn-like** appendages that are **prominent in some species,** and probably help **to** keep **the** parts of

CHAPTER XV

the flower in place despite the struggles of entrapped guests. I have always found it easy to capture bees on the flowers, and have never failed to find traces of pollen masses on their legs; very often only the little discs remain, the pollinia having been severed.

The cactus is a type of plants that, without dying down to the ground, can adapt themselves to a long dry season. Mexico and the arid regions of America generally, have the greater part of all the known species of cactus, but South Africa has many similar plants belonging to the genera Ficoideæ, Crassulaceæ, Portulacaceæ and the like. Plants of this type are found also on the faces of rocks where there is very little soil, and on sea sands where evaporation is rapid. In order to present the least possible evaporating surface, in all these plants cylindrical or spherical forms are approached. In the cacti it is the stems that assume these forms, but in century plants, the Crassulaceæ (live-for-ever, hen and chickens, and the like), ice plants, &c., it is the leaves. These plants must always have an epidermis that restricts evaporation, either a very thick one, as in the cactus, or one with a waxen or silicious coating. They have also what is termed aqueous tissue, i. e., tissue that stores water. In the cactus, this tissue is in the interior, its cells and their contents being transparent; in many of the plants, the aqueous tissue is visible beneath the epidermis as transparent lines or dots. The ice plant, Mesembryanthemum, common on our beaches and also cultivated in gardens, has water stored in crystalline vesicles that thickly cover stem and leaves; it is not fully understood why in this last case the water in the delicate-walled vesicles does not evaporate easily; it is probably because of dissolved salts in the cells and the nature of the cell walls. Where these plants are exposed to the attacks of animals, silicious coats or dissolved mineral salts sometimes serve defensive purposes, but the cactus, like the yucca and century plant, has savage weapons. Our common tuna cactus, *Opuntia Lindheimeri*, var. *occidentalis*, has not only the long spines, which are modified leaves, but also at the bases of the spines numbers of maddening, little, barbed bristles; one of these bristles under the microscope is represented in Fig. 16. It is not unusual for one cactus to have three kinds of weapons.

For a few weeks in spring time, the tuna cactus has true leaves, like little fleshy horns, but most cacti have only the spines. Fragments of the tuna take root very readily; they are easily detached from the parent plant, and the spines must aid in their distribution. Probably there are few seedlings. The flowers of the tuna open but once, and then for a few hours only, but they are thronged with bees that jostle one another in their eagerness for pollen; the bees usually,

SUPPLEMENT

though not always, strike the prominent stigmas on entering, and while the stigma stands sometimes above the anthers when the flower is expanded, it is probably self pollinated as the flower closes. I have seen the stamens resume their vertical position in from eight to twelve minutes, but they have rarely opportunity to do this on account of the frequency of guests. The rarer, but more showy, magenta cactus, *Opuntia basilaris*, Engl., has more active stamens, and a stigma that is always quite above them. The smaller greenish flowers of *Cereus Emoryi*, Englm., are so narrow that a large bee must cling to the ample stigma in order to collect the pollen, and so he is sure to effect cross pollination.

Collections of cacti are so common in Southern California that many other interesting features may be observed. The night-blooming Cereus climbs by branching rootlets; the expansion of its great exquisite flowers is a process well worth watching. Two species of Mesembryanthemum are common on our beaches. Two genera of the Crassulaceæ, Sedum and Cotyledon, are rather widely distributed in rocky places; the species are difficult to determine. The chamisal, *Adenostoma fasciculatum*, H. & A., has its short, spine-like leaves in fascicles or clusters. Like most Rosaceæ, it provides honey, and is much cross pollinated by bees although its flowers seem quite capable of self pollination.

The Eriogonums have already received considerable attention in Chapters III and XIV. Their pollination is interesting; the following is true of *E. fasciculatum* and of several other species with large flower clusters. The perianth consists of six sepals; the nine stamens erect themselves, a few at a time, holding their anthers, which are of contrasting color, upward during dehiscence; after the anthers have all shed their pollen and fallen off, the three styles, which have been tightly curled down in the centre, straighten and hold their stigmas in the same position before occupied by the anthers. The perianths do not fall after the pollination of the flowers, but change to a contrasting color, those of the white species to pink, the yellow ones to red, and so add to the attractiveness of the cluster. The amount of honey secreted is sufficient to attract throngs of bees of all sizes, although they must share it with all sorts of small insects.

The dodder, Cuscuta, is parasitic on many of our native plants; there are several species, difficult to determine. The little flowers are somewhat fragrant and have considerable honey; at first their anthers are held away from the stigmas, but later on the stamens bend to the centre, and apply their pollen to the stigmas.

Pentstemon cordifolius, Benth., the Pentstemon with cordate

CHAPTER XV

(heart-shaped) leaves, is common in the south, and is well worth watching, from the time it sends out new shoots after the first rains, as noted at length in the Supplement to Chapter V, until it unfolds its brilliant flowers in early summer. It is an excellent example of weaving plants, that is of plants that clamber up through more woody underbrush; the different disposition of the flowers as well as of the leaves on the branches, which are sometimes vertical, sometimes horizontal, looped or pendent, are interesting subjects for out of door study. The pollination of this and other scarlet Pentstemons mentioned later on among mountain flowers, is, on the whole, similar to that of the scarlet Pentstemon noted in Chapter XI, but in the details there are interesting differences that can be easily discerned in field work. The anthers and stigmas of Pentstemons always lie against the upper lip, or roof, of the flower and in most scarlet species well out to the tip, where they are sure to be struck by the bird's bill or head as he enters; by a downward curve of the style, the stigma is so placed that it is struck first.

There are several species of large, rose, violet or purple-hued Godetias that bloom in May and June, the most common species in the vicinity of Los Angeles being *G. Bottæ*, Spach., the one of the illustration. Little tufts of hairs near the base of the petals form a ring about the style, and exclude the less desirable guests. Dehiscence of pollen goes on very slowly, and the four-lobed stigma does not unfold and expose the stigmatic surface until the pollen is nearly or quite shed. In newly opened flowers, the style keeps the closed stigma out of the way, sometimes quite outside the corolla, but by executing an elaborate movement it finally places the exposed stigma in the way of entering guests. *Clarkia elegans*, Lindl., of the same illlustration, is rather local in the south, but is more abundant and attains greater size in the north.

The Indian pink illustrated in the Reader, is *Silene laciniata*, Cav.; it is said to be less handsome than its northern cousin *S. Californica*, but it is a brilliant posy, and is sought out by humming birds, although the plants do not grow in masses. As in the Pentstemon, anthers and stigmas, when mature, are above the entrance to the honey, and situated where they will strike the bird's head. The stamens spread out fan-like, a few anthers dehiscing at a time; finally, the styles lengthen and put the stigmas in the place of the now fallen anthers. The wild fuchsia, *Zauschneria Californica*, is another example of a choripetalous flower made exclusive by united sepals. The pollination of this flower has been already described in the Supplement to Chapter III.

119

SUPPLEMENT

Many summer plants in the valleys are, as noted in the Reader, Compositæ, and have been considered in Chapter XII and its Supplement. The Trichostema (blue-curls), and Eremocarpus (turkey-weed) were treated at length in Chapter III and Supplement. These plants can, of course, depend upon much insect attention; even the inconspicuous flowers of the Eremocarpus and other Euphorbiaceæ are much visited by bees.

It is difficult to dismiss the subject of mountain flowers with a few sentences. All through our California mountains at an altitude of from five to eight thousand feet, are green meadows or pine groves traversed by mountain streams, spots ideal in scenery and climate, and yet unspoiled by civilization, where those who enjoy a natural life can, at little expense, pass a delightful summer. There are parents who appreciate this, and fortunate children who will enjoy all the more their weeks of liberty if they have seeing eyes for living things. There are teachers, too, who, even in this high pressure age of summer schools, institutes, seminaries, and pedagogical literature, believe that a few weeks spent with nature in searching out her lessons at first hand, is better preparation for another year's work than continuous inpourings of pedagogical lore.

A complete summary of mountain plants is really not necessary. One must turn back the leaves of the calendar in climbing mountains, and expect, at an altitude of seven thousand feet, to find early violets and buttercups in June. The species will be unfamiliar and it may not be possible to learn the full name without consulting some botanist who has made a special study of the neighborhood, but any one interested in the habits of plants and their adaptation to their environment, will take keen pleasure in comparing the mountain species with familiar valley species of the same genus. Generally, too, the plants grow in such masses that there is the best of opportunities for watching their relations to their guests. Then there is the zest of finding plants of new families; early in the season Saxifrages, with their delicate flower clusters and beautiful leaves, also the tall, blue iris, or flag; later, rhododendrons and gentians, and perhaps orchids and cardinal flowers; in the pine woods, not only the snow plant, *Sarcodes sanguinea*, Torr., but other saprophytic Ericaceæ, the pine drops, *Pterospora andromedea*, Nutt., for instance, with its tall, uncanny spikes of flesh-colored flowers, and the pretty Pyrolas, which are only partially saprophytic. There are plants whose flower clusters actually melt their way through the snow; the Soldanella of the Alps does this, but so far as I can learn, our snow plant has not this habit. It is possible that red flowers are not so marked a feature of

CHAPTER XV

the mountain flora in the north, **but in the San Bernardino mountains** there are literally acres of Pentstemon and **Castilleia at altitudes of** seven or eight thousand feet, and scattered **scarlet flowers everywhere.** For two months in **the** mountains, I **daily watched humming** birds visiting flowers from **dawn to** twilight, **and I never once saw** them visit any **but scarlet flowers. In the** valleys, **they frequently** avail themselves of **the hospitality of flowers of other colors, but where** there is **a great abundance of** flowers of all colors, **their behavior** leaves one **in little doubt as to** their color preference.

The methods **of climbing plants are** full **of interest, and form the subject of** considerable botanical literature ; **they can be easily studied in the home garden during the** summer **vacation. Four** different **types have already** been noted among native plants ; **the** *Pentstemon cordifolius*, **the blackberry and** the *Nemophila aurita*, climb by weaving or looping themselves over underbrush ; **the** morning-glory and **dodder climb by means of** their **twining** stems, the chilicothe **and grape, by tendrils, and the poison oak by** rootlets. Encourage the children to find further illustrations of these types among cultivated plants. **The climbing roses** weave **their way up on lattices, just as** the Nemophila or **the** blackberry **clambers over bushes, the prickles** always assisting in the process. The best way to study twining **plants** is to watch morning-glory seedlings when the upper portion **of the** young shoot begins its sweeping motion, before it has found a **support.** In this way it will be seen that there actually is motion, not merely **growth,** around the support. This motion is called circumnutation ; **it is not twisting, but the stem** bends successively to all points of the compass. **The** books **explain that** circumnutation is due **to a line of** turgescence, that is, of much distended cells, which by moving about the stem, causes it to bend always in the opposite direction ; **but what** impels the protoplasm that controls the **turgescence cannot be** explained. **A very vigorous morning-glory stem upon a warm** day completes a **revolution in two or three hours,** so the motion may be easily **perceived.** Twining plants need slender supports, and in temperate **climates it is usually** only annuals that choose this **method of getting up in the world quickly.** When a perennial twines about a perennial stem either **one or the other** must be ultimately destroyed, since **as the** supporting stem **increases** in circumference, either it must be strangled, or the twining **stem** ruptured. Let children observe whether the direction of the twining is clockwise or the reverse. I know but one plant, the so-called smilax, that twines sometimes in one, sometimes in the other way. The best device of all for climbing is by means **of** tendrils, and here again there **is much** diversity of

SUPPLEMENT

method. The tendril may simply form a loop or ring about the support, like the nasturtium; or it may circumnutate and coil after grasping the support, like the passion vine; or it may turn away from the light in seeking attachment, like the Virginia creeper and other plants accustomed to climb up walls, rocks, or tree trunks. In the nasturtium it is the leaf or flower stem that acts as a tendril; in the pea, modified leaflets; in the grape, chilicothe and many others, modified branches; and so on. Let children find out for themselves the habits of the tendrils of the sweet pea, grape, passion flower, chilicothe, pumpkin, &c. At first the tendrils are long and straight; if they must push their way through under-brush, they do this after the manner of weaving plants, keeping compactly together; once free to seek a support, they circumnutate until a free tip, often hooked, catches some object; at once the tendril coils, and perhaps by this means brings its neighbors, also, to some support; tendrils not reaching an object, coil and usually shrivel and drop. The coiling of the tendrils not only brings the plant nearer to its support, but on account of the resulting elasticity, renders it less likely to be torn away by storms. Light-avoiding tendrils are sometimes provided with little discs; more often the tendril insinuates itself into some minute crevice or hollow, and then its tip develops a disc-like attachment that moulds itself into the crevice like wax, completely filling it. The English ivy, so common in cultivation, climbs by aerial rootlets. Let the children find out whether these rootlets come only at joints or all along the stem, and whether on the side toward, or away from, the light. Do these roots feed the plant also, as do the suckers of the dodder? Let them find this out by cutting the twining stem so that it has no connection with the soil. As noted before, the leaves of climbing plants spread themselves out fully to the light, often forming beautiful mosaics.

Beach plants have the same devices and the same general appearance as desert plants and those active during the dry season. This has been set forth at length in Chapter III. Some individual beach plants have also been noted, the sand primrose in Chapter IX; the beach lupine in Chapter X; sand verbenas in Chapter XIV; and succulent beach plants in this Chapter.

CHAPTER XVI.

WEEDS.

The great extent of foreign plant immigration is perhaps sufficiently emphasized in the Reader. To begin at once, then, with the individual weeds :—The malva, bur-clover, and filaree have already been much considered in the Reader. The fox-tail grass, *Hordeum murinum*, Linn., is surely familiar to all Californians. Both its awns, and its inner floral envelopes, which are sharp pointed at the base, are provided with bristles, rendering the plant an almost universal pest. Prof. Hilgard of the State University, in his report for the year 1890, is my authority for the statement that the soft brome grass, *Bromus mollis*, Linn., is likely to aid in the extirpation of the fox-tail grass; this brome grass is not yet abundant in the south. The wild oat, *Avena fatua*, Linn., is valued for hay, although its awns sometimes do mischief to cattle and horses. An extremely pretty foreign grass, *Achyrodes (Lamarckia) aureum*, O. Ktze., with fluffy, plume-like spikes of minute flowers, is becoming very common in the south. It seems unfair to include this pretty grass among weeds, but it probably has little value for pasturage, and it is well provided with means of dispersal. It is doubtless true that the Bermuda grass will never be troublesome on land not irrigated, but it ruins many a lawn. The chess, though frequently imported with grain seed, does not succeed in California, but the darnel, a species of Lolium, is troublesome in some parts of the state. The alkali grass, *Distichlis maritima*, Raf., is not easily exterminated from moist alkali soil.

The extermination of black mustard, probably introduced as a condiment, is the most serious weed problem of the south. Such is the vitality of its multitudinous seeds, that mowing the flower stalks for a single season seems to have but little permanent effect. The seeds are utilized commercially to some extent. The radish, *Raphanus sativus*, Linn., is a rather common weed, but is not generally persistent.

SUPPLEMENT

The yellow melilot, *Melilotus Indica*, All., with **fragrant clover-like leaves, and** slender spikes of minute yellow flowers, is very common **in the** south, but it does not infest **grain fields as in** the north. *Melilotus **alba**, Lam*, the very fragrant **melilot with white** flowers, is not yet common. The water cress, *Nasturtium officinale*, R. Br., **is** not an unmixed evil. The **course** of **introduced weeds is** usually, like the star of empire, westward; but the pond **weed**, *Elodia Canadensis*, referred to in the Reader, is a notable exception. **The** chilicothe, already noted **in** Chapter V, is very troublesome to fruit growers **in** Antelope **V**alley and other reclaimed regions of scanty moisture. A near relative **of** the chilicothe, *Cucurbita fœtidissima*, H B **K., known** as mock orange, Chili-cojote, or calabazilla, frequently spreads itself over waste ground and neglected fields in the **south,** and very troublesome **it is** if it gains a foothold in cultivated **land.** It **has** an underground part of enormous extent, and its stems **with** their ill scented foliage trail off five or six feet in every direction; its yellow, gourd-like fruits are exceedingly numerous, and the seeds are well protected by the hard rind of the gourd.

Poison oak persists for some time in newly cleared land, new shoots springing up from fragments of its sturdy rootstocks. For the same reason, bracken is difficult to eradicate **from its former habitat, and** several years are necessary **to free moist alkali soil from such** plants as the yerba **mansa (Chapter XIV), wild heliotrope** (Chapter IX), wild celery **and other** swamp-loving Umbelliferæ. **It** is natural that **plants with** underground parts that enable them **to** stand repeated croppings, should survive in pasture lands that are little cultivated. The sanicles and other Umbelliferæ mentioned in the Supplement to Chapter XII, are examples of this, also the pretty blue-eyed **grass,** *Sisyrinchium bellum*, Fig. 61. The numerous little corms given off each season by *Brodiæa capitata* (Chapter VII) render it quite persistent, though not troublesome, **in** grain fields. In sandy soils *Lupinus formosus* develops very long roots that are difficult to exterminate. The potency of the morning-glory as a weed also resides in its underground part, and in this case the outcome of the struggle between the weed and the agriculturists is by no means certain. In the south it is the foreign morning-glory, *Convolvulus arvensis*, Linn., that sometimes drives the farmer from the field, but Prof. Hilgard states that in the bay region it is the native *C. Californicus* that remains in possession of **abandoned** orchards. The pre**eminence of** Amsinckia and Eschscholtzia among native weeds is **certainly due to their ability to endure** drought, and it is in the **regions where rainfall is particularly uncertain** that these two weeds flourish. When several dry years come in succession, it is interesting to watch the struggle for supremacy between the two weeds; for a time the Amsinckia, whose qualifications are set forth in the Supplement to Chapter IX, take the lead, but ultimately the perennial plant crowds the annual to the wall.

Of the minor annual weeds of winter and spring time, *Orthocarpus purpurascens*, illustrated in Fig. 54, the well known tidy-tips, and *Phacelia tanacetifolia*, Fig. 45, and its near kin, are simply vigorous natives that do not at once succumb to cultivation. The two former are mildly troublesome in grain fields, the last is common along waysides and in waste places generally. The Calandrinias, Chapter IX, Supplement, are short-lived, but their fleshy leaves and numerous seeds are elements of strength. Our native nettles flourish all the year if left in their chosen habitats. The dodder is sometimes very troublesome in alfalfa fields. *Tropidocarpum gracile*, Hook., is a yellow-flowered Crucifer rather persistent in irrigated loamy soils, especially in the desert. Another Crucifer, the shepherd's-purse (Chapter IX), makes itself at home in California as everywhere else in the world, but can hardly be termed troublesome. Of introduced weeds of the pink family, *Silene gallica*, Linn., is very prolific in some warm sandy soils; it is nearly as viscid as its cousin *S. laciniata* of Chapter XV, and has short, one-sided clusters of white or pinkish flowers of medium size; the spurry, *Spergula arvensis*, requires more moisture, and is even more local; the chickweed, *Stellaria media*, as noted in Chapter IX, is widely distributed and persists the year round. The pimpernel, noted in Chapter XIV, Supplement, is another not very offensive weed that blooms and fruits the year round.

Many of the summer and autumn weeds have been already considered; the milkweed, Asclepias, in the previous chapter; Trichostema and Eremocarpus, in Chapter V and elsewhere; and, in the Supplement to Chapter XII, the various weeds belonging to the Compositæ,—the sunflower, tar-weeds, ragweeds (Ambrosia), fleabane (Erigeron), Mayweed (*Anthemis cotula*), cockle-bur (Xanthium), sow-thistle (Sonchus), and Spanish-needle (Bidens.) The sow-thistle is illustrated in the Reader; the cockle-bur and Spanish-needle, like the hoarhound, are too notorious to need illustration; the Mayweed seems to be rapidly gaining ground in the north; in the south a nearly related native plant, *Matricaria discoidea*, DC., is more aggressive, but neither is yet troublesome. The yellow star thistle, *Centaurea melitensis*, Linn., is a pest that perhaps ranks next to mustard and wild barley in the north, but in the south it is limited to pasture lands. The two species of Euphorbia, *E. albomarginata* and *E.*

SUPPLEMENT

polycarpa, known as rattlesnake weed, and described in Chapter XIV, have some very weedy habits, although on the whole they are more attractive than troublesome; they are supposed to provide a remedy for rattlesnake bites. The Euphorbia that is by some believed to be poisonous to gophers, is *E. Lathyrus*, Linn., an escape from cultivation. The skunkweed is *Gilia squarrosa*, H. & A., more common northward; *Portulaca oleracea*, the purslane, so troublesome in our Eastern States, is only occasional in California, although it seeks to adapt itself to the dry season by reducing its fleshy leaves to cylinders. Many of our most familiar weeds belong to the nearly related families, Polygonaceæ, Amarantaceæ, and Chenopodiaceæ. The most common species of Chenopodium are foreigners, and are known as pigweed or goosefoot; the one in the illustration is *C. Album*, Linn. Atriplex is another genus of Chenopodiaceæ, and *A. bracteosa*, Wats., is a common tumble-weed along Los Angeles streets. The very common tumble-weeds of cultivated lands however, are species of Amaranthus; they are too familiar to need description. Tumble-weeds are generally supposed to scatter their seeds as they are rolled about by the wind, but in this section the seeds fall before the plants begin to roll; and, like most of this group, they may well trust the dispersal of their minute and very numerous seeds to the wind. The doves eat great quantities of these seeds, and may aid in distributing them. Rumex, the dock, and Polygonum the knotweed, are genera belonging to the same family as the Eriogonum ; fragments of these plants are illustrated in the Reader. The *Polygonum aviculare*, Linn., of the picture, spreads out in flat circular mats two or three feet in diameter, attaining in summer, on our hard, baked adobe soil, a much greater size than in the Eastern States. Its slender, woody stems are green, and must perform much of the vegetative work ; it is said that cattle will eat it when hard pressed. All these plants have small and inconspicuous flowers, and at first glance one would pronounce them self pollinated, as most of them doubtless are.; but the flowers are exceedingly numerous, and the wind probably aids considerably in their pollination ; in fact, on closer examination, one finds that several species are quite incapable of self pollination, either because they are unisexual or because they mature their stamens only after their own stigmas are past the receptive stage. As stated in Chapter III, it is not easy to explain fully the success of these ubiquitous weeds, which thrive in the most unpromising places. They seem to have inherited virility from their ancestors, which, for centuries, and in many countries, have been conquering weeds.

There are in Southern California several really ornamental plants

CHAPTER XVI

which are commonly classed with weeds. The "Jimson" weed is properly *Datura stramonium*, a common weed in the Atlantic States, which took its name from the Jamestown of colonial days. This weed is occasionally seen in California, but its common name is applied to our native *Datura meteloides*, DC., which has the same poisonous and narcotic qualities. The native Datura is a handsome plant, despite its rather ill repute and disagreeable foliage; its huge white or violet-tinted flowers are produced in profusion from May to November. The fennel, with its exceedingly delicate foliage, grows so luxuriantly during the summer and autumn about the towns of Southern California that stories of fennel thickets being resorts for wild beasts in Asia Minor seem quite credible. The castor-oil plant, *Ricinus communis*, Linn., generally cultivated in gardens and greenhouses, grows spontaneously in the south, frequently attaining the height of trees; and the tree-tobacco, *Nicotiana glauca*, Graham, becomes a graceful tree in two or three years. These two plants speedily transform unsightly vacant city lots of southern towns into miniature parks. Some peculiarities of the leaves of the Ricinus were noted in Chapter III. The flowers are unisexual, the pistillate flowers terminating the clusters, which bear staminate flowers lower down. The stamens are clustered like bunches of grapes, and produce a great quantity of fine, dry pollen; the stigmas, too, are characteristic of wind-pollinated flowers, being very large and rough. Bees visit the plants for the viscid substance contained in the glands previously mentioned, but they are not likely to play any part in pollination. The seeds and their distribution were considered in Chapter II. The tree-tobacco is a recent immigrant from South America. Like other species of tobacco, it produces almost incredibly numerous seeds; they are held in open capsules, and are carried far and wide on the winds, so that young seedlings are seen springing up everywhere, even on perpendicular walls of clay and on the ruins of old adobe structures. The trees are in flower all the year round, and in even the coldest weather, humming birds can count on their hospitality, for the flowers, although yellow in color, are in form and structure well adapted to the birds; they are pendent, long and tubular, and are rendered still more exclusive by incurving stamens; they secrete honey abundantly, and attract throngs of birds. If these were native flowers, we should consider the bearing of their color and their guests upon the theories of flower coloration, since all our natives of similar structure are scarlet; but we should need first to know of the insect and bird life in the native home of the tree tobacco; so, like the birds, we are content to accept them and enjoy them without question.

APPENDIX.

SUGGESTIONS FOR THE USE OF THE READER IN THE SCHOOL ROOM.

In this day of laboratories, the use of a Botanical Reader may be questioned. The aims of this book are stated at length in the Preface, and it may be taken as an apology for the book's existence, if apology be necessary. The writer has endeavored to arrange the book so that it cannot be used as a substitute for field and observation work, but only as a stimulus and aid. She has found California teachers generally very willing to undertake real nature study; but teachers who must unfold all subjects to many children in limited time, can not be specialists, and need every legitimate aid in work of this kind. It is easy to lead children to open their eyes to natural objects and to induce them to bring specimens for school-room study, but their expression of what they see shows how much their observations need to be interpreted and supplemented. Of course the Reader should not be used until children have found out what they can for themselves, and have expressed it in their own way. The order of the chapters conforms to the changing seasons in our climate, beginning in the autumn, but it would not be wise to attempt the entire book in one year. The early chapters were written for children in the fourth and fifth grades, but some of the later chapters should not be undertaken before the seventh or eighth grades. Courses of nature study are usually made to extend over the entire eight primary and grammar grades, and this is doubtless best if the work can be sufficiently varied to keep up keen interest. It has been thought best to suggest in detail the adaptation of the Reader to a course of plant study, although the index of plants and topics at the end of the book renders it easily adaptable to any course. The course selected is similar to the one now followed in the Los Angeles City Schools.

APPENDIX

FIRST GRADE.

AUTUMN.—Castor-bean, morning-glory and pine seedlings. Develop the fact that the seed is a little plant, also, in a general way, the use of root, stem and leaves.

WINTER.—Study, out of doors if possible, pine tree, mature castor-oil plant and growing morning-glory, noting most prominent features only. In school room, note pollen and teach its use; study pine cones and castor-oil fruits, that is, seed protection and seed distribution.

SPRING.—Note morning-glory flower and fruits and develop the use of the flower with its color and honey.

The aim of the work of the first three years should be to teach leading facts about entire, living plants. The ideal plan is to have the plants growing in school gardens, but the seedlings can be grown in window gardens, tended by the children. If there is no pine tree or castor-oil plant in the vicinity of the school building, some other tree with abundance of pollen may be substituted, or the study of the willow—see work of next grade—may be taken up instead. In work with the flower here, teach only that the flower helps in seed-making by inducing insects (or birds) to carry pollen. Have the children find honey and pollen in other flowers growing out of doors near the school building, and let them watch for the flowers' guests. In the first three grades do not dissect flowers; think of them always as living wholes. For further suggestions, see index, under pine, castor-oil plant, etc.

SECOND GRADE.

AUTUMN.—Other dicotyledenous seedlings suggested in Chapter II. Volunteer seedlings from children's home gardens, such as "nasturtium" (Tropœolum), sweet pea and geranium (Pelargonium), transplanted into school garden. Study of growth from perennial underground stems, children contributing to the school garden calla rootstocks, Chinese lily bulbs, and, perhaps, underground storehouses of iris, canna, hyacinth and the like. Obtain if possible Mariposa lily bulbs and a California poppy growing from a strong rootstock.

WINTER.—Watch growth from buds, from shoots kept in water in the school room if out of door observation is not feasible. Study willow (see Chapter VIII). Also, if it seems best, some of the other native trees, the sycamore, cottonwood or alder. Teach stigma and its use. Note seed distribution of the trees under consideration.

SPRING.— Using the flowers of "nasturtium," geranium, sweet pea, calla, Chinese lily, poppy, Mariposa and iris, have the children find

APPENDIX

pollen, stigmas and honey if any Note honey guides in nasturtium, some Pelargoniums and iris. Develop fact that insects may seek shelter as well as honey. Watch for the guests of the flowers in home or school gardens; they can often be seen in passing along city streets.

THIRD GRADE.

AUTUMN.—Corn seedlings, with experiments suggested in Chapter II, to show uses of root-hairs, root-tips, and woody strands.

WINTER AND SPRING.—Further study of trees. Select some accessible nut-bearing tree, walnut is best, and a tree with fleshy fruit, peach, orange or apple for instance. Watch development from flower to fruit and study seed distribution. (See " nuts " and " fleshy fruits, " index.) Teach names of trees common in the vicinity of the school, and note their most striking characteristics;—deciduous or evergreen, pollinated by wind or insects, method of seed distribution, etc. Collect and study as many native plants as possible, considering the flowers as wholes, as suggested in first grade work; also noting seed distribution. Talk of advantages of flowers in clusters and note that sunflower, thistle, tidy-tips, marguerite and the like, are flower clusters.

FIELD WORK.—It should be possible in this grade for children to take occasional excursions to hillsides or cañons, bringing their collections to the schoolroom. It would be well to keep on the blackboard a growing list of the native plants the children know. In Southern California such a list at the end of the third grade should include a considerable number of the following plants ;—willow, walnut, sycamore, oak, cottonwood, alder, elder, poison-oak, "scrub" oak, wild blackberry, wild grape, wild lilac, wild currant and gooseberry, California holly, some ferns, poppy, shooting star, violet, lupine, cluster lily, Mariposa lily, buttercup, cream-cup, mustard, painted cup, monkey-flower, blue-eyes, ground-pink, nightshade, four-o'clock, tidy-tips, sunflower and thistle. Encourage children to make collections of seeds (or fruits) with floaters, hooks, etc.

FOURTH GRADE.

AUTUMN.—Central idea, plants as food-makers. Collect and study Algæ, then take up Chapter I, Reader. Drill on new terms chlorophyll, oxygen, carbonic acid gas, protoplasm, cell. It is believed that the subject of nutrition is more easily taught from lower plants and that the novelty of the material will give zest to the subject. Read also Chapter II, preceding it by a study of any of the seedlings not

APPENDIX

hitherto considered, and emphasizing the physiological part. Have children grow mould on bread.

WINTER.—Have children collect and study lichens and toadstools and read Chapter IV.

SPRING.—Study of a flower in detail, the peony if accessible, otherwise the buttercup. Read first part of Chapter VII. Drill in pointing out parts of other flowers collected.

FIELD WORK.—Continue third grade work, but let pupils keep individual lists with dates, continuing to add to them during the summer vacation. Have common wayside plants, such as bur-clover and filaree included.

FIFTH GRADE.

Main topic for the year's work, climate and vegetation.

AUTUMN.—Have children collect summer and autumn plants, study them with experiments, then read Chapter III. Emphasize transpiration current, and the necessity for maintaining the balance between water supply and evaporation.

WINTER —Chapter V, preceded by preparatory work. Collect and study ferns; read Chapter VI.

SPRING.—Study fig and other common trees of the vicinity not previously taken up. Read Chapter VIII. Observation work with grasses, wild oats as a type; also with palms, calla, canna and banana. Correlate this work closely with Geography. With the chapter on trees, talk of trees of different climates; let grasses suggest grains, sugar cane and bamboo; the fig, India rubber; the canna, the tropical aroids, and so on. Dwell on the palms of economic importance and teach something of the plants that yield cotton, tea, coffee, spices, tapioca and the like. Consult the Supplement to Chapters XIII and XIV using index. If only local geography is taught in this grade, defer this work and take that suggested in the next grade.

FIELD WORK.—Continue the collecting and the lists with dates, adding to the lists and comparing dates with dates of previous years.

SIXTH GRADE.

Review of plant physiology and more detailed study of pollination of flowers.

AUTUMN.—Review physiological parts of Chapters I, II, III, and V.

WINTER AND SPRING.—Study pollination of some of the following flowers ;—four-o'clock, poppy, shooting star, violet, cluster lily, Mariposa lily, nemophilas, Gilias, nightshade, lupine, Mimulus, Collinsia, painted cup, owl's clover, Pentstemon, larkspur, cactus,

APPENDIX

wild fuchsia and blue-curls. Have children notice the relative positions of stigmas and anthers and their time of maturity; also any special devices for excluding undesirable guests, or protecting pollen or honey, or for securing self pollination. Read all of Chapter VII and parts of Chapters IX, X, XI and XV.

FIELD WORK.—Watch insects on the above flowers. Have children note the most common weeds, continuing observations during summer.

SEVENTH GRADE.

AUTUMN.—Compare notes on weeds and read Chapter XVI.

WINTER AND SPRING.—Pollination of as many of the following as seems feasible:—*Phacelia tanacetifolia*, alfalfa, filaree, geranium, sages, sunflower, thistle and milkweed, finishing **Chapters IX, X, XI, XII and XV**. In connection with this work note family characteristics so that at the end of this grade children will be able to recognize members of most of the following families:—lily, poppy, mustard, pea, evening-primrose, Gilia, blue-eyes, forget-me-not, Scropulariaceæ, Labiatæ and Compositæ.

FIELD WORK.—Out of door study of the above. Collecting plants for a herbarium is an excellent stimulus for field work, and this may sometimes be practicable in grammar grades. The amount of time available for this line of work must be considered. At least urge contributions of specimens from children who spend the summer at the seaside or in the mountains.

EIGHTH GRADE.

Review of Reader with some classification in mind. Give more attention to reproduction. Possibly a herbarium of typical plants might be required.

AUTUMN.—Chapters I and IV, considering as many species of Algæ and Fungi as seems feasible.

WINTER.—Ferns and Coniferæ, Chapter VI and first part of Chapter XIII.

SPRING.—Endogens and Exogens, Chapters XIII and XIV.

The work outlined in each grade assumes that the work of the previous grades has been completed. Of course the classification suggested for the seventh and eighth grades should not be undertaken unless there is already an acquaintance with a large number of plants. Often these last two years can be best used for picking up loose ends. The author considers the work outlined in the first three and the fifth grades of most importance.

INDEX

This index includes the plants and topics of both Reader and Supplement, the numbers following "s" referring to the Supplement.

Abies, s 100.
Abronia, s 110.
Abyssinia, s 115.
Acacia, 108, 180, 181, s 23, 38, 63, 113.
Acanthus, s 114.
Acer, see Maple.
Achillea, s 93.
Achyrodes, s 123.
Acorn, 31, Fig. 8, p. 31, s 13, 61.
Adenostoma, s 118. See also Greasewood.
Adiantum, 77, 79, Fig. 27, p. 76. s 46. See also Maiden-Hair.
Africa, South, s 113, 117.
Agaricus, s 34.
Agave, s. 103.
Aggregatæ, 182.
Akene, s 53.
Albumen, s 5, 12, 19.
Alder, 40, 71, 104, 172, Fig. 38, p. 103, s 28, 41, 59, 60, 109.
Alfalfa, 129-132, 193, Fig 50, p. 130, s 78, 175.
Algæ, 17-24, 56, s 3, 10-13.
Alkali Grass, s 105, 123.
Alkali Weed, 174, Fig. 64, p. 173. See also Yerba Mansa.
Allium, s 56. See also Onion.

Allspice, 179.
Almond, 180.
Alnus, see Alder.
Alyssum, Sweet, 175, s 68.
Amanita, s 34.
Amarantaceæ, s 110, 126.
Amaranthus, s 126.
Amaryllis Family, 163-164.
Ambrosia, s 91.
Amelanchier, s 113.
Amsinckia, 121, 122, 207, s 72, 123.
Anemopsis, 174, Fig. 64, p. 173, s 109.
Angiosperms, 162, s 101.
Anise, 157.
Annuals, 39.
Anthemis, s 93.
Anther, Definition of, 87.
Anther, s. 49.
Antheridium, s 10, 42-45.
Anthocyanin, s. 41.
Antirrhinum, s 85.
Apetalæ, s 108, 110.
Apium, s 95.
Aplopappus, s 25, 90.
Apple, 107, 179, s 59, 62.
Apricot, 105, 107.
Apron Kelp, s 9.
Aqueous Tissue, s 117.

INDEX

Arabia, s 115.
Arbor Vitæ, s 98.
Arbutus, 70, 181, s 40, 65.
Archegoniatæ, s 42.
Archegonium, s 42-45.
Arctic Regions, 54.
Arctostaphylos, s 65.
Argemone, s 67.
Aristolochiaceæ, s 109, 110.
Aroids, s 104.
Artemisia, s 25, 93.
Asarum, s 110.
Asclepias, 186-189, Fig. 66, p. 188.
   s 116.
Ash, s 110.
Asparagus, 164.
Aspidium, s 46. See also Shield Fern.
Asplenium, s 46.
Aster, s 25, 89, 90.
Astragalus, 80.
Audibertia, 146-150, Fig. 57, p. 149.
   s 86.
Australia, 43, 174, 179, s 109.
Autumn Leaves, 48, s 27, 28.
Autumn Plants, 39-50, s 20, 28.
Avena, s 126. See also Wild Oats.
Awns, s 106.
Azalea, 181, s 113.
Azolla, s 47.

Baby-Blue-Eyes, 117, s 70, 71.
Baccharis, s 90.
Bachelor's Button, s 94.
Bacteria, 51-52, s 17, 29, 30.
Bæria, s 76, 91.
Balsam, s 81.
Bamboo, 168, s 105.
Banana, 162.
Banyan, s 109.
Barberry, s 32, 111.
Bark, s 58, 59.

Barley, 162, 168, s 106. See also Foxtail Grass.
Basswood, s 112.
Bast, s 58.
Bay, 175.
Beach Plants, 41.
Bean, 29, 133, Fig. 7, p. 30, s 13.
Beech, 172, s 109.
Bees, 47, 70, 71, 89, 90, 92, 95, 100,
   101, 108, 112, 116, 119, 121, 122,
   123, 125, 128, 129, 131, 133, 140,
   142, 144, 146, 148, 150, 155, 184,
   186, 188, 194, 197, 198, s 41, 51-
   55, 60, 64-67, 70-74, 78, 87, 96,
   102, 116-118, 120, 121, 127.
Beet, 170, s 110.
Beetles, 57, 90, 110, s 34.
Beggar's Ticks, s 95.
Begonia, 175.
Bermuda Grass, 170, 205.
Beta, s 110.
Bidens, s 21, 92, 95.
Big Chia, 146, 148, s 86.
Bigelovia, s 25.
Big-Root, see Chilicothe.
"Big Tree," 109, 161, s 61, 99.
Bilabiate Families, 138-150, s 83-88.
Birch, 172, s 109.
Bird-Foot Fern, 79 80, Fig. 28, p.
   78, s 46.
Birds and Seeds, s 63.
Bitter Clover, 205, Fig. 72, p. 206
Blackberry, 108, 179, s 62, 75, 121.
Black Lily, s. 104.
Black Sage, 146-148, Fig. 67, p.
   149. See also Sages.
Bleeding Heart, s 111.
Bloomeria, 163, s 102.
Blue-Curls, 47, 150, 197, 201, 207,
   Fig. 15, p. 49, s 26, 86. See also
   Trichostema.
Blue-Eyed Grass. 181, 207, Fig.

INDEX

61, p. 165, s 76, 123.
Blue-Eyes Family, 117-121, 181.
Blue Flowers and Bees, 184, 198 s 51.
Boraginaceæ, s 69, 114.
Bowlesia, s 95.
Box Elder, s 112.
Bracken, 80, 198, 201, s 46, 124.
Bracts, Leaves among Flowers.
Brass Buttons, s 93.
Brassica, s 68. See also Mustard.
Brazil-Nut, 179.
Bread Fruit, 174, s 109.
Breathing, 41. See also Respiration.
Breathing Pores, 36, 37, Fig. 10, p. 36. See also Stomata.
Brodiæa, 91-93, Fig. 33, p. 91, s 57-53, 101, 123. See also Cluster Lily.
Brome Grass (Bromus), 205, s 123.
Broom, California or Wild, 40, 44, Fig. 11, p. 45, s 24, 80. See also Lotus glaber.
Broom-Corn, s 106.
Brown Algæ, 21, 22, Fig. 2, p. 20, s 8.
Brown Lily, 163, s 56.
Brussels Sprouts, 176.
Bryophytes, s 42
Buckeye, 177, s 112.
Buckwheat, Wild, 44, 174, 191, Fig. 12, p. 45, Fig. 68, p. 192, s 24, 110, 118. See also Eriogonum.
Buds, 71, 100, 104, Fig. 38, p. 103, s 41, 57, 59.
Bulb, 64-66, 93, 183, Fig. 22, p. 65.
Bulrush, 164.
Bumble-Bee, 129, 133, 148, 171, s 78.
Bunch Grass, 168, s 106.
Bur-Clover, 39, 61-62, 130, 132, 202,

Fig. 20, p. 62, Fig. 50, p. 131, s 37, 38, 78.
Bur-Marigold, s 21, 92.
Burdock, 211.
Buttercup, 89, 90, 97, 198, Fig. 32, p. 88, s 53, 110.
Buttercup Family (Ranunculaceæ), 175.
Butterfly, 92, 93, 116, 121, 131, 142, 155, 186, 194.
Butterfly Tulip, 125, Fig. 48, p. 124, s 74.
Button Bush, s 115.
Button Sage, 146, 148. See also Black Sage.

Cabbage, 175, 176.
Cactus, 42, 44, 175, 176, 178, 189-191, 201, Fig. 16, p. 50, Fig. 67, p. 190, s 24, 117.
Calabazilla, s 124.
Calandrinia, 75, 201, 207, s 110, 125.
Calico Kelp, s 9.
Calla, 97 98, 162, 164, Fig. 36, p. 98, s 54, 104.
Calochortus, 123, 125, Fig. 48, p. 124, s 73, 74. See also Mariposa.
Calycifloræ, 178.
Calyx, Definition of, 87.
Cambium, s 59, 101.
Camellia, s 112.
Campanulaceæ, s 114.
Camphor, 175, s 111.
Camphor Weed, s 26. See also Trichostema.
Canada Thistle, 211.
Canaigre, 200, s 110.
Cancer Root, s 83.
Canna, 160, 170, s 106.
Canterbury Bells, s 114.
Canterbury Bells, see Phacelia Whitlavia.

INDEX

Caper Family, s 111.
Capparidaceæ, s 111.
Caprifoliaceæ, s 115.
Capsella, s 68, 69.
Capsule, s 53, 54.
Caraway, 157, s 94, 95.
Carbo-hydrates, s 5.
Carbonic Acid Gas, 19, 37, 52, s 5, 18.
Cardinal Flower, s 114, 120.
Carduus, s 93.
Carnation Pink, 174.
Carrot, s 94, 95.
Carum, s 95.
Caryophyllaceæ, s 110.
Cascara, s 65.
Cassia, 175, s 111.
Castilleia, 142, 144, 200, Fig. 54, p. 143, s 85, 120.
Castor-Oil Plant, 25-27, 36, 38, 117, 211, Fig. 5, p. 26, Fig. 10, p. 36, s 11, 26, 127.
Casuarinaceæ, s 109.
Catalpa, s 114.
Catasetum, s 107.
Catch-Fly, s 110.
Catkin, 100.
Cat-Tail, 162, 164, 170.
Cat's Ears, 125.
Caulicle, 32.
Cauliflower, 176.
Ceanothus, s 40, 64. See also Lilac.
Cedar, s 98.
Celery, 157, s 94, 95.
Cell, 18, s 3, 4.
Cell Division, s 6.
Cellular Structure, 35-37, Fig. 10, p. 36, s 9, 17, 58.
Cellulose, s 5, 19.
Centaurea, 210, Fig. 72, p. 106, s 94, 95, 125.

Century Plant, 162, 164, s 24, 103, 117.
Cercocarpus, s 65.
Cereus, s 118.
Ceylon, s 111.
Chænactis, s 92.
Chamisal, 191, s 118.
Chamomile, s 96.
Chaparral, 109, 191, s 80.
Cheeses, 75, 176. See also Malva.
Cheilanthes, s 46.
Chenopodiaceæ, s 110, 126.
Chenopodium, s 126.
Cherry, Wild, 107, 109, 179, 180, s 65.
Chess, 211, s 125.
Chestnut, 172, s 99, 109.
Chia, 146, s 86.
Chickweed, 94, 174, 202, s 55, 110, 125.
Chicory, 150, s 94.
Chilicothe, 66-68, 182, 201, 207, Fig. 23, p. 67, s 39, 114, 121, 122, 124.
Chili Cojote, s 124.
Chili Pepper, s 73.
Chinese Lily, 99, 164, s 54.
Chlorogalum, 183, 184, s 39, 116. See also Soap-Root.
Chlorophyll, 19, 24, 37, 51, s 4, 22.
Chloroplasts, s 4.
Chocolate, 176, s 112.
Choripetalæ (Petals not united), s 110.
Chorizanthe, s 110.
Chrysanthemum, s 93.
Cinchona, s 115.
Cinnamon, 175, s 111.
Citrus, s 112.
Cladophora, s 6, 8.
Clarkia, 194, Fig. 69, p. 195, s 119.
Classification, 160, s 83, 98.

INDEX

Clematis, 40, 175, s 21, 111.
Climbing Nemophila, 117, 119, Fig. 44, p. 117.
Climbing Pentstemon, 194-196, 197, Fig. 75, p. 196, s 41, 121.
Climbing Plants, 68, 117, 119, 123, 133, 193, 196, 199, s 38, 40, 41, 80, 82, 119, 120-121.
Clocks, see Filaree.
Close Pollination, see Pollination.
Clover, 132, 133, s 38, 78, 79.
Cloves, 179.
Club-Moss, s 43, 48.
Cluster Lily, 66, 90 93, 97, Fig. 22, p. 65, Fig. 33, p. 91, s 53, 54.
Clusters of Flowers, see Flower Cluster.
Cnicus, Fig. 59, p. 156, s 93. See also Thistle.
Coal and Ferns, 84.
Cocklebur, 182, 202, 210, s 91.
Cocoa-Nut Palm, s 105.
Cocometa, 66. See also Cluster Lily.
Coffee, s 115.
Coffee, California, s 65.
Coffee Fern, 78, 79, Fig. 27, p. 76, s 46.
Collinsia, 140-142, Fig. 53, p. 141, s 84.
Color of Autumn Leaves, 48, s 28.
Color, Physiology of, s 51.
Columbine, 175, 197.
Communities of Flowers, 155, 159.
Competition among Plants, 50.
Compositæ, 44, 151-159, 182, 197, 200, s 25, 75, 89-97, 114.
Compound Pistil, 92, s 49.
Conductive Tissue, 35, 37, s 17, 58.
Cone, 27, 161, Fig. 7, p. 28, Fig. 40, p. 106, s 61.
Cone-Bearers, see Coniferæ.

Coniferæ, 27-29, 161, 162, s 23, 44, 61, 98, 100.
Convolvulaceæ, s 73, 114.
Conium, s 94.
Convolvulus, see Morning-Glory.
Copa de Oro, s 66.
Coprinus, s 34.
Corallina, s 9.
Cordylanthus, s 86.
Coreopsis, s 92.
Corethrogyne, s 95.
Coriander, 157, s 94, 95.
Coriandrum, s 95.
Cork, s 58.
Cork-Oak, s 58, 108.
Corm (a Solid Bulb), s 53.
Corn, Indian, 32, 38, 42, 162, 168, 170, Fig. 9, p. 32, s 13, 105.
Cornaceæ, s 113.
Corolla, Definition of, 87.
Cotton, 176, s 112.
Cottonwood, 40, 71, 104, Fig. 50, p. 103, s 28, 60, 108. See also Poplar.
Cotula, s 93.
Cotyledons, 27-34, s 118. See also Seed-Leaves.
Coulter's Poppy, 112, s 67.
Cow-Tree, 174.
Crabs and Sea Mosses, 22.
Cranberries, 181.
Crape Myrtle, 179.
Crassulaceæ, s 117, 118.
Cream-Cup, 112, Fig. 41, p. 111, s 66, 67.
Crimson-Flowered Sage, 150.
Crocus, 162, 164.
Cross-Pollination, see Pollination.
Croton, Hairs of, Fig. 16, p. 50.
Crown Imperial, s 101.
Cruciferæ (Mustard Family), 175-176, 207, s 57, 68, 125.

INDEX

Crude Sap, 37.
Cryptograms, s 44.
Cucumber, Wild, see Chilicothe.
Cucumber Vine (Cucurbita), 182, s 124.
Cucumber Family (Cucurbitaceæ) 182, s 114.
Cup-Fungus, s 35.
Currants, Wild, 68-70, 108, 178, Fig. 24, p. 69, s 40, 64.
Cuticle, s 17, 23.
Cuscuta, s 73, 118. See also Dodder.
Cyclamen, 181.
Cycads, s 98.
Cypress, 108, 109, Fig. 40, p. 106, s 98.
Cypripedium, s 106.

Daisy, 151, 154.
Dandelion, 151, 154, 155, s 94.
Dangers to Plants, 41-42.
Date Palm, s 105.
Datura, s 22, 73, 114, 121. See also "Jimson" Weed.
Darlingtonia, s 111, 112.
Darnel, s 123.
Daucus, s 95.
Deciduous, 48, s 27, 28.
Decay and Bacteria, 52.
Decaying Leaves, 48.
Delphinium, s 116. See Larkspur.
Dendromecon, s 67.
Desert Plants, 181, s 22. See also Dry Season Plants.
Desmid, s 57.
Diatom, s 7.
Dicentra, s 111.
Dicotyledons, 172-182, s 101, 108, 115.
Dill, 157, s 94.
Dimorphic, s 70, 72.

Diœcious (Staminate and pistillate flowers on different plants), s 96.
Dipsaceæ, s 114.
Disease and Bacteria, 61, s 29, 30.
Distribution of Corms, s 53.
Distribution of Seeds, see Seed Distribution.
Distribution of Spores, 57, s 34.
Distribution of Willows, s 59.
Ditch Grass, s 104.
Divided Leaves, 64, 75.
Dock, 170, 202, 210, Fig. 73, p. 209, s 110, 126.
Dodder, 181, 191-194, Fig. 68, p. 196, s 73, 118, 121, 125.
Dodecatheon, 95, 97, Fig. 35, p, 96, s 54.
Dogbane, s 114.
Dogwood, 113.
Dormant Buds, 57.
Douglass Spruce, 100.
Dragon Tree, 102.
Drosera, 111, 112
Drupe (Stone Fruit,) 62.
Dry Climate, Plants of, s 20-28.
Dry Rot, s 36.
Dry Season, Plants of, 39-50, s 22-27.
Duckweed, 166, s. 104.
Dust and Plants, 41, s 22.
Dutchman's Pipe, s 109.

Earth Star, 57-59, Fig. 19, p. 58, s 35.
Ebony, 181.
Egg-Cell, s 10, 42, 50.
Egg Plant, s 73, 113.
Eidelweiss, s 90.
Elder, s 28, 61, 115
Ellisia, 121, Fig. 45, p. 118, s 71.
Elm, 172, s 109.
Elodea, 207, s 124.

INDEX

Embryo, 32.
Embryo Sac, s 11.
Emmenanthe, s 71.
Encino, s 108.
Endogens, 160-171, 98-107.
Endosperm, 32, s 12, 13.
Encelia, s 91.
English Ivy, s 113, 122.
Epidermis (skin), 42, s 17, 23.
Epiphytes, s 107.
Equisetum, 83, Fig. 29, p. 82, s 43. 47.
Eremocarpus, 41, 47, Fig. 14, p. 47, s 22, 25, 120. See also Turkey-Weed.
Ericaceæ, s 120. See also Heath Family.
Erigeron, s 90.
Eriodictyon, s 72.
Eriogonum, 44, 174, 191-193, Fig. 12, p. 45, Fig. 68, p. 192, s 24, 110, 118. See also Buckwheat.
Eriophyllum, s 92.
Erodium, 134-136, s 38. See also Filaree.
Erythræa, s 114.
Erysimum, s 68.
Eschscholtzia, 110-112; Fig. 41, p. 111, s 66, 133. See also Poppy.
Eucalyptus, 42, 43, 107, 179, s 23, 54, 99.
Euphorbia, 177, 178, s 126.
Euphorbiaceæ, 177-178, s 112.
Evaporation, see also Transpiration.
Evening Primrose, 178, 179, s 69, 70.
Evening-Primrose Family (Onagraceæ), 114, s 69, 119.
Everlasting Flower (Gnaphalium), 44, 46, 68, s 25, 90.
Exogens, 172-182, s 101, 108-115.
Explosive Fruits, s 81.

Fairy's Lantern, 123, Fig. 48, p. 124.
Falling of Leaves, s 27, 28.
Families of Plants, 160-182, s 89-115.
Feather Moss, s 9.
Fennel, 157, 202, s 94, 95, 96, 127.
Ferns, 72-84, s 37, 45-47.
Ferns not Eaten, s 47.
Fern Prothallium, 74, Fig. 30, p. 84.
Fern "Seeds," 73, 74, s 47.
Ferns, Young Plants, 74, 75.
Fertilization (Union of sperm and egg cells), 87, s 50. See also Pollination.
Fertilizing Cells, s 42.
Ficoideæ, s 117.
Ficus, s 109. See Fig.
Fig, 107, 172, 174, Fig. 38, p. 103, s 63, 109.
Filament, Definition of, 87.
Filaree, 39, 61-64, 135-136, 202-204, Fig. 20, p 62, Fig. 21, p. 63, Fig. 51, p. 134, s 14, 38, 80, 81.
Fimbriaria, s 44.
Firs, 109, s 60, 100.
Flag, 99, 162, 164, s 120. See Iris.
Flax, 177, s 112.
Flea-Weed, s 26. See Turkey-Weed.
Fleshy Fruits, 107, 180, s 62, 63, 113.
Fleshy Plants, 42, 44, s 20, 24.
Flies, 57, 110, 121, 154, 186, s 34, 35, 39, 53, 74, 82.
Flower Cluster, 97, s 96.
Flowering Maple, 176.
Flowering Willow, 90.
Flowers, Structure and Function of, 85-89, s 49-50.
Fly-Flower, s 76.
Fœniculum, s 95.

INDEX

Food in Seeds, 27, 29, 39, 180, 203.
Food-Making, 17-24, 29, 43, 62, s 4-6, 18.
Forget-me-not, 114, 121, Fig. 46, p. 120, s 69.
Forget-me-not Family (Borginaceæ), 121, 181, s 69, 72, 73.
Four-O'clock, 174, s 55, 110.
Foxglove, s 114.
Foxtail Grass. 61, 170, 202, 204, 205, Fig. 20, p. 62, s 123.
Franseria, s 91, 95.
Freemontia, s 65.
Freesia, 164.
Fritillaria, s 56.
Fuchsia Family, see Evening-Primrose Family.
Fuchsia, Wild, 48, 49, 194, 195, 197, Fig. 15, p. 49, s 26, 119. See also Zauschneria.
Fucus, s 8.
Fungi, 52-59, s 29-36.

Galium, s 115.
Gamopetalæ, s 113.
Gentian, 182, 198, s 114, 120.
Geraniaceæ (Geranium Family), 135-137, s 80-82.
"Geranium," Cultivated, 136-137, 177, s 80, 81.
Germination of Seeds, 25-38, s 11-19.
Gigartina, s 9.
Gilia, 114-117, 181, 198, 207. Fig. 43, p. 115, s 52, 69, 70, 126.
Gilia Family (Polemoniaceæ), 181.
Ginger Plant, 170.
Gladiolus, 162, 164.
Glandular Hairs, s 26.
Globe Tulip, 123, Fig. 48, p. 124, s 74.
Gloxinia, s 114.

Gnetaceæ, s 98.
Gnaphalium, Fig. 16, p. 60, s 25, 90. See also Everlasting Flower.
Godetia, 194, Fig. 69, p. 195, s 119.
Golden-Back Fern, 75, 77, Fig. 27, p. 76, s 46.
Golden Ear-Drop, s 111.
Golden Fields, 154, s 76.
Golden Lily Bell. 123.
Golden-Rod, s 25, 90.
Golden Star Lily, 163, s 102.
Golden Yarrow, s 92.
Gold Thread, 59, 193. See Dodder.
Gooseberry, 68, 70, 108, 178, Fig. 25, p. 70, s 40, 64.
Goosefoot, s 110, 126.
Grape, 40, 177. s 21, 112, 121, 122.
Grasses, 61, 137, 162, 168-170, s 105, 106.
Greasewood, 109, 180, 191. See Adenostoma.
Green Algæ, 17-21, Fig. 4, p. 23, s 3-7.
Grevillea, 179.
Grindelia, s 89.
Ground Cherry, s 73.
Ground Pink, 116, Fig. 43, p. 115, s 70.
Groundsel, s 76, 91, 92.
Guava, 179.
Gum Arabic, 181.
Gymnogramme, s 46.
Gymnosperms, 161-162, s 98.

Habenaria, 176, s 106.
Hairs, Plant, 41, 44, 46, 47, Fig. 16, p. 50, s 25.
Heath Family (Ericaceæ), 181, s 113.
Heather, 181.
Hedge-Nettle, s 86.
Helianthus, 151-154, Fig. 58, p. 125, s 25, 91. See also Sunflower.

INDEX

Heliotrope Family, 121, s 69.
Heliotrope, Wild (Heliotropium), s 72, 124.
Heliotrope, Wild (Phacelia), s 71, 125.
Hemizonia, s 92, 93.
Hemlock, 157.
Hemp, s 109.
Hen-and-Chickens, s 20, 117.
Heteromeles, s 22, 65.
Heterotheca, s 89.
Hibiscus, 176.
Hickory, s 109.
Higher Plants, 25.
Hoarhound, 42, 68, 146, 202, 210, s 25, 86, 114.
Holdfast, 21, 22, Fig. 2, p. 20.
Holly, California, 40, s 22, 65, 113.
Holly, English, or True, s 112.
Hollyhock, 75.
Honey, see Pollination.
Honey Guides, 93, s 53, 55, 82.
Honeysuckle, s 65, 115.
Hop, s 109.
Hordeum, s 123. See also Foxtail Grass.
Horse-Chestnut, 177, s 112.
Horse-Radish, 176.
Horsetail, 83, Fig. 29, p. 82, s 43, 47.
Hosackia, 133, s 24. See also Lotus.
House-Fly Fungus, 53.
Houttuynia, s 109.
Hoya, s 114.
Huckleberry, 181, s 113.
Humboldt Lily, 163.
Humid Climate, Trees of, s 58.
Humming Birds, 48, 70, 108, 109, 137, 139, 140, 142, 144, 150, 155, 186, 194, 196, 198, s 27, 40, 51-51, 64, 81, 82, 84, 85, 86, 119, 121, 127.
Hyacinth, 164.

Hybrids, s 87-88.
Hydrocotyle, s 94.
Hydrodictyon, 17-21, Fig. 1, p. 18, s 3.
Hydrophyllaceæ, s 64, 114.
Hygrometric, 83, s 35, 46.

Ice Plant, 170, s 117.
Impatiens, s 81.
Incense Cedar, s 98.
India, 179, s 111, 112.
Indian Lettuce, s 55.
Indian Pink, 174, 194, 197, Fig. 70, p. 195, s 119. See also Silene.
Indian Plume, 142, Fig. 54, p. 143. See also Castilleia.
Indian Warrior, s 86.
India-Rubber, 174, 178.
Indigo, s 113.
Inferior Ovary, 113, 163, 178.
Injurious Fungi, 53, 54.
Innocence, s 84.
Insectivorous Plants, s 111.
Insects and Color, s 51-52.
Insects and Plants, see Pollination.
Iris, 99, s 55, 120.
Iris Family, 164.
Involucre (around flower clusters), s 97.
Isomeris, s 111.
Ithuriel's Spear, s 101.
Ivory Nuts, s 105.
Ivy, s 113, 122.
Ivy-Geranium, s 81.

Jack-in-the-Pulpit, s 104.
Jamaica, 83, s 45.
Japan, 164.
Jasmine, 182.
Java, s 109, 115.
"Jimson" Weed, 181, 210, s 22, 73, 127. See also Datura.

INDEX

Joint-Firs, s 98.
Jonquil, 164.
Judas Tree, 180.
Juglans, see Walnut.
Juncaceæ, see Rushes.
Juniper, s 98.

Kail, 176.
Kelp, 21, 22, Fig. 2, p. 20, s 80.
Kentucky Blue Grass, s 105.
Knotweed, 174, 210, Fig. 73, p. 209, s 110, 126.
Kryuitzkia, 72.

Labiatæ, 146-150, s 39, 86-88.
Lace Fern, 80, Fig. 28, p. 78, s 46.
Lace Pod, s 68.
Lady Fern, s 46.
Lady Slipper, s 106.
Lady Washington, s 81.
Lamarkia, s 123.
Larkspur, 75, 175, 184, 186, 197, Fig. 65, p. 185, s 111, 116.
Lathyrus, s 79.
Laurel, 175, s 61, 111.
Lavatera, s 65.
Lavender, s 114.
Layia, s 75, 91.
Leaf Arrangement, s 30.
Leaf-Bud of Ferns, s 47.
Leaf Mosaic, 68, s 38.
Leaf Movement, 62-64, s 39, 82.
Leaf Protection, 70, 75-80, s 22, 23.
Leaf, Structure of, 37, 37, s 17.
Leaves and Dew, s 41.
Leaves and Rainfall, 97, 99.
Leaves of Autumn Plants, 40-41.
Leaves of Endogens, 162.
Leaves of Exogens, 172.
Leaves, Sleeping Position of, s 38.
Leguminosæ, 126-133, 180, 181, s 30, 77-80, 113.

Lemnaceæ, s 104.
Lemonade Bush, s 64.
Lemon Seeds, s 13.
Lemon Verbena, s 114.
Leopard Lily, 163.
Lepidium, s 68.
Lepidospartum, s 24.
Leptosyne, s 76, 91, 92.
Lettuce, s 94.
Lettuce, Miner's, s 55.
Libocedrus, s 98.
Lichen, 54 57, 161, Fig. 18, p. 55, s 32-34, 103.
Light and Plants, 19, 37, 41, 62-64, s 22.
Liguliforæ, s 94.
Lilac, California, or Wild, 70, 108, 177, 182, s 40.
Liliaceæ (Lily Family), 162-164, s 56, 101.
Lily, 162, 163, 164, 198, 207.
Lily-of-the-Valley, s 101.
Lime, s 112.
Linaria, s 85.
Linden, s 112.
Little Chia, 146-148, s 86. See also Salvia.
Live-for-ever, 178, s 20, 117.
Live Oak, 43, 107, s 23. See also Oak.
Liverworts, 81, Fig. 30, p. 84, s 37, 42-45.
Lobelia, s 114.
Lobularia, s 68.
Loco Weed, 133, s 80.
Locust, s 113.
Lolium, s 123.
Lonicera, s 65.
Lotus, 133, Fig. 11, p. 45, s 24, 80. See also Broom.
Love-Vine, see Dodder.
Lunularia, s 44.

INDEX

Lupine, 68, 126-129, 140, 142, Fig. 21, p. 63, Fig. 49, p. 127, s 40, 77-78, 123.
Lycopodium, s 48.

Mace, 175, s 111.
Macrocystis, 22, Fig. 2, p. 20, s 8.
Macrospore, s 43.
Madia, s 92.
Madrone, 109, 181, s 65, 113.
Magnolia, 175, s 23, 59, 111.
Mahogany, California, 107, 180, s 64.
Maiden-Hair Fern, 77, 79, Fig. 27, p. 76, s 46.
Malacothrix, 154, Fig. 13, p. 46, s 25, 94.
Malva, 61, 62, 175, 176, 202, 203, Fig. 20, p. 62, s 37.
Malvaceæ (Malva Family), 175, 176, s 65, 112.
Malvastrum, s 65.
Manioc, 178.
Man-Root, see Chilicothe.
Manzanita, 43, 108, 109, 181, s 23, 65, 113.
Maple, 177, s 61, 112.
Marine Algæ, 21-24, s 7-10.
Mariposa, 123, 125, 194, 200, Fig. 48, p. 124, s 73-74.
Marrubium, s 25.
Marsh Pennywort, s 94.
Maté, s 112.
Matilija Poppy, 112, s 67.
Matricaria, s 125.
Mayweed, 211, s 93, 125.
Meadow Rue, 175, s 111.
Meconopsis, 112, s 67.
Medicago, 129-132, Fig. 50, p. 130, s 78.
Medullary Rays, s 58.
Megarrhiza, see Chilicothe.

Melilotus, 205, Fig. 72, p. 206, s 124.
Melon, 182, s 114.
Mentha, s 86.
Mentzelia, Hairs of, Fig. 16, p. 50.
Myrsiphyllum, s 101.
Mesembryanthemum, s 117, 118.
Metabolism, s 6, 18, 19.
Mexico, 129, 189, s 117.
Micrampelis, 66, 68, Fig. 23, p. 67, s 39.
Micromeria, s 86.
Micropyle, s 11.
Microspore, s 43.
Mignonette, s 111.
Milkweed, 182, 186-189, Fig. 66, p. 187, s 114, 116.
Milkweed (Sonchus), see Sow-Thistle.
Mimulus, 138-140, 150, 197, Fig. 52, p. 139, s 21, 83-84.
Minerals Used by Plants, 35, 48, s 16.
Miner's Lettuce, 97, s 55, 110.
Mint, 146, 200, s 86.
Mint Family (Labiatæ), 146-150.
Mirabilis, s 55, 110.
Mistletoe, s 109, 110.
Mock Orange, s 124.
Mock Willow, s 90.
Mohave Desert, s 66, 70.
Moisture-Loving Plants, s 37-41.
Monardella, s 86.
Monocotyledons, 160-171, s 98-107.
Monopetalæ, s 113.
Monterey Cypress, 161, s 61, 98.
Monterey Pine, s 12, 61, 100.
Morning-Glory, 28, 114, 123, 181, 193, 201, 207, 211, Fig. 7, p. 30, s 13, 73, 121, 123.
Mosses, 81-83, s 37, 43, 45.
Moths, 137, 139, 140, 142, 184, 189, s 27, 69, 70, 81, 84, 85, 102, 103.

INDEX

Mould, 52-53, Fig. 17, p. 53, s 30, 31.
Mountain Flowers, 197, 198.
Mountain Mahogany, s 65.
Mountain Phlox, or Pink, 116, 117, Fig. 43, p. 116, s 70.
Mountain Plants, s 120-121.
Mountain Trees, 109.
Mouse Ear, s 74.
Movements of Roots, s 16.
Movements of Sap, s 18, 19.
Muilla, s 56.
Mulberry, 172, 174, s 109.
Mullein, s 85.
Mushroom, 59, s 34-35.
Mustard, 112, 113, 200, 202, s 68, 111, 123.
Mustard Family, 113, 175, 176, s 67-68.
Mustard Seeds, s 14.
Mycelium, s 31, 32, 34.

Naiadaceæ, s 104.
Naked-Seeded Plants, 161, 162. See also Gymnosperms.
Nasturtium (Water Cress), s 28, 124.
Nasturtium (Tropæolum), 29, 35, s 13, 80, 81, 82, 122.
Native Weeds, 201.
Nelumbium, s 111.
Nemophila, 117, 119, 200, Fig. 44, p. 117, s 69, 70, 71, 121.
Nereocystis, s 8.
Nettle, 174, 210, s 32, 109, 125.
Nicotiana, s 73, 127.
Night-Blooming Cereus, 189, s 118.
Nightshade, 40, 114, 122, 123, 181, Fig. 47, p 122, s 21, 69, 73.
Nightshade Family (Solanaceæ), 181, s 69.
Nitrogen, 35, 52, s 5, 17.
Norway, 161.

Notholœna, s 46.
Nucleus, s 4.
Nutmeg, 175, s 111.
Nut-Pine, s 100.
Nuts, Protection and Distribution of, 28, 105.
Nyctaginaceæ, s 110.

Oak, 172, s 59, 60, 61. See also Live Oak.
Oats, Wild, 168-170, Fig. 63, p. 169, s 106, 123.
Odor of Plants, 41.
Œnothera, 114, Fig. 42, p. 113, s 69.
Oil, 27, 38, s 12.
Okra, 176.
Oleander, s 114.
Olive, 182, s 114.
Onagraceæ, see Evening-Primrose Family.
Onion, 32, 34, 163, 164, s 56.
Oögonium, s 10.
Opuntia, 189, 191, Fig. 67, p. 190, s 117, 118.
Opium, s 111.
Orange, 108, 177.
Orchard Trees, s 62.
Orchids, 171, s 107, 120.
Oregon Pine, s 100.
Organic Substances, 19, 35-38, s 5, 19.
Orthocarpus, 142, 207, Fig. 54, p. 143, s 54, 85, 125.
Oscillatoria, s 7.
Ovary, Definition of, 85.
Ovule, Definition of, 85, s 11, 44.
Owl's Clover, s 84, 85, 125.
Oxalis, s 38, 80, 81.
Oxygen, 19, 24, 37, s 4-5, 18, 29.

Pæonia, see Peony.

INDEX

Painted Cup, 142, 196, 197, Fig. 54, p. 143, s 85.
Painter's Brush, 142, Fig. 54, p. 143, s 84, 85.
Palisade Cells, s 17, 23.
Palm, 162, 166, 168, s 104-105.
Palm Seeds, 34.
Pansy, Yellow, see Violet.
Papaver, s 67.
Papaveraceæ, s 66-67.
Papilionaceæ, s 78, 84.
Papyrus, 170.
Paraguay Tea, s 112.
Parasite, 51-60, 193. See also Fungi.
Parry's Lily, 163.
Parsley, s 94.
Parsnip, s 95.
Passion Flower, 179.
Pastinaca, s 95.
Paulownia, s 114.
Pea, 31, 133, Fig. 7, p. 30, s 13, 79, 122.
Peach, 105-107, 179, Fig. 39, p. 105, s 62.
Pea Family, 126-133, 179-181, s 77-88.
Peanut, 29, 31, 38, 133, s 113.
Pear, 105-107, 179, s 62.
Peat, 83.
Pecan, 172.
Pectocarya, s 72.
Pedicel, Flower-Stem.
Pedicularis, s 86.
Pellæa, s 46.
Pelargonium, s 80, 81. See also Geranium.
Pennyroyal, s 86.
Pentstemon, 146, 150, 194, 198, Fig. 55, p. 147, Fig. 71, p. 196, s 41, 51, 52, 85, 118, 119, 120, 121.
Peony, 85-89, 97, 175, Fig. 31, p. 86, s 49, 52, 110.

Pepper, 174.
Pepper Grass, 175, s 68.
Pepper Tree, 108, s 23, 59.
Periwinkle, s 114.
Persimmon, 181.
Petal, Definition of, 87, s 49.
Petal, a modified Stamen, s 49.
Petiole, Leaf Stem.
Petunia, s 73, 114.
Peucedanum, Fig. 60, p. 158, s 95.
Peziza, s 33.
Phacelia, 114, 119, 121, 181, Fig. 45, p. 118, s 32, 68, 71, 125.
Photo-Synthesis, s 5, 18.
Phyllotaxy, s 38.
Physalis, s 73.
Physiology of Seedlings, s 14-19.
Pie-Plant s 110.
Pigments, s 5.
Pigweed, 202, 210, Fig. 73, p. 209, s 110, 126.
Pimpernel, 181, s 202, 113, 125.
Pine, 27-29, 107, 108, 109, 161, 162, Fig. 40, p. 106, s 60, 100.
Pine-Apple, 170, s 103.
Pine Flowers, s 61.
Pine Seeds 27-29, Fig 7, p. 28, s 12.
Pink Family, 110, 125.
Pink Painter's Brush, s 84, 85. See also Orthocarpus.
Piñon, 28, s 12, 100.
Pinus, s 61, 100.
Pistil, Definition of, 85.
Pistil, Modified Leaf, s 49.
Pitcher Plant, 176, s 111, 112.
Pith, s 101.
Plagiobothrys, Fig. 46, p. 120, s 72.
Platanus, see Sycamore.
Plantain, 202, s 107.
Plant, Products, 37.
Platystemon, s 67.

INDEX

Platystigma (Little White Poppy), 112, s 67.
Plocamium, 22, Fig. 3, p. 23, s 9.
Plum, 179, s 62.
Plumbago, s 113.
Plumule, 32.
Poinsettia, 177.
Poison Hemlock, s 94.
Poison Oak, 40, 48, 68, 108, 177, 186, 201, s 21, 28, 64, 121, 124.
Poisonous Plants, 25, 157, 178, s 11, 21, 22, 34, 35, 73, 80, 94, 109, 114, 126.
Polemoniaceæ (Gilia Family), s 69, 114.
Pollen, Definition of, 87.
Pollination, conveyance of pollen to the stigma.
    Close, of the same flower.
    Cross, of another flower.
    Self, without foreign aid.
    68, 70, 71, 87, 88-112, 114-159, 164-171, 180-199, s 26, 27, 39-41, 50, 52-55, 60-75, 78-87, 91, 95, 96, 102, 103, 106, 113, 116-121, 126, 127.
Polygonaceæ, s 110, 126.
Polypetalæ, s 110.
Polypodium, 72, Fig. 26, p. 73, s 45.
Polyporus, s 35, 36.
Pome (Apple-like Fruit), s 62.
Pomegranate, 179.
Pond Scum, 17-21, s 6.
Pond Weed, 207, s 104, 124.
Pop-Corn Flower, s 73.
Poplar, 104, Fig. 38, p. 103, s 108. See also Cottonwood.
Poppy, 110-112, 175, 200, 201, 207, Fig. 41, p. 111, s 66, 111, 124.
Poppy Family, 112, s 66.

Populus, s 60, 108. See also Poplar and Cottonwood.
Pore-Fungus, see Polyporus.
Pores, see Stomata.
Portulaca, 174, 207, s 75, 126.
Portulacaceæ, s 110, 117.
Potato, s 73, 113.
Potato Family, (Solanaceæ), s 69. See also Nightshade Family.
Potato Fungus, 53.
Potentilla, 75.
Prickles, 41.
Prickly Pear, 189. See also Cactus
Prickly Poppy, 112, s 67.
Primrose, Fig. 42, p. 113, 181, 198, s 113.
Progress of Weeds, 202.
Protection of Plants, 41-50.
Protection of Pollen, 87, 90, 92, 93 etc. See also Pollination.
Prothallium, 74, s 43.
Protophytes, s 7.
Protoplasm, 19, 37, s 4.
Protoplast, 21.
Prunus, s 65. See also Cherry.
Pteridophytes, s 42.
Pteris, s 46.
Pterospora, s 120.
Puff-Ball, 35, 59, s 85.
Pumpkin, 182, s 114.
Purslane, 174, s 110, 126.
Pussy Ears, s 74.
Putrefaction, s 30.
Pyrola, s 120.

Quince, 107, 178, s 62, 65.
Quercus, see Oak.
Quinine, s 115.

Radicle, 32.
Radish, 175, 176, s 68, 123.
Rafflesiaceæ, s 109, 110.
Ragweed, 211, s 91.

INDEX

Rainy-Season Plants, 61-71, s 37-41.
Ranunculaceæ, 175, s 53, 110.
Raphanus, s 68, 123.
Raspberry, 179, s 62.
Rattle Pods, 133, s 80.
Rattlesnake Weed, 177, s 126.
Receptacle, 135.
Red Algæ, 22-24, s 8-9.
Red Flowers, 197, 198. See also Humming Birds.
Red-Hot Poker, 164.
Red Pigment, 2?, s 8, 17, 41, 59.
Red Poppy, 112.
Redwood, 109, 161, s 99.
Reindeer Moss, 54.
Reproduction, 21, 22, 51-59, 85-87, s 6, 9-10, 29, 31, 33, 41-44, 50.
Resin Weed, s 89.
Respiration, s, 5, 15, 22.
Resurrection Plant, s 48.
Rhamnus, s 65.
Rhododendron, 181, s 113, 120.
Rhubarb, 174, s 110.
Rhus, 177, s 64, 112.
Ribes, s 40, 64.
Rice, 168, s 106.
Ricinus, s 127. See also Castor-Oil Plant.
Richardia, s 104. See also Calla.
Roble, s 108.
Rock Fern, 72. See also Polypodium.
Rockweed, 21, 22, s 8.
Rolled Leaves, s 24.
Romneya, s 67.
Root, 32, 34-35, s 14-16.
Root-Hairs, 34-35, s 14-15.
Root Pressure, s 18.
Roots, Pulling, s 53.
Rootstock, 65-68, Fig. 22, p. 65, s 54. See also Storehouses.
Root-Tips, 34-35, s 14-16.

Rosaceæ (Rose Family); 179, 180, s 62, 113, 116.
Rose, 179, 180, s 87, 121.
Rose Bay, s 113.
Rose Hip, s 113.
Rosemary, s 114.
Rosewood, s 113.
Rubiaceæ, s 115.
Rubus, see Blackberry.
Rumex, s 126.
Rushes, 162, 164, 170, s 103.
Russian Thistle, 211, s 23, 110.
Rust, 53, s 31, 32.
Rye, s 106.

Sage, 146-150, 191, Fig. 57, p. 149.
Sage-Brush, 44, 68, s 25, 93.
Sago Palm, s 105.
Salix, see Willow.
Salsify, s 94.
Salvia, 146-148, Fig. 56, p. 147, s 86. See also Chia.
Sambucus, s 61.
Sand Lupine, 201, 207.
Sand Spurry, s 110.
Sand Verbena, s 110.
Sanicle (Sanicula), s 75, 95, 96, 123.
Sap, 37, s 17.
Sap, Conduction of, s 58.
Saprophytes, 51-60, 198, s 107, 113, 120. See also Fungi.
Sarcodes, s 120. See also Snow Plant.
Sassafras, 175, s 111.
Satin Bell, 123, Fig. 48, p. 124.
Saxifrage, 178, 198, s 75, 120.
Scaly Fern, 80.
Scarlet Flowers, 19, 198, s 51-52. See also Humming Birds.
Scarlet Poppy s 67.
Scouring Rush, 83, Fig. 29, p. 82, s 47.

INDEX

Scrophularia, s 52, 85.
Scrophulariaceæ, s 83-85.
Scrub Oak, s 65.
Scutellaria s 75, 86.
Sea-Lettuce, s 8.
Sea-Mosses, 21-24, s 7-10.
Sedum, 42, 44, 178, s 20, 118.
Seed, 160.
Seed Distribution, 25, 27, 60, 62, 97, 102, 104, 105, 122, 129, 135-137, 154, 155, 159, 170, 180, 188, 189, 203, 210, 211, s 11, 53, 54, 59, 60-63, 67, 72, 74, 78, 80, 81, 87, 91, 92, 95, 96, 106, 107, 110, 126, 127.
Seed-Leaves, 27-34, 61. See also Cotyledons.
Seedlings, 25-38, 61, 193, Fig. 20, p. 62, s 11-19.
Seed-Making, 87.
Seed Protection, 25, 27. See also Seed Distribution.
Sedges, 168, 170.
Selaginella, s 43, 44, 48.
Senecio, s 22, 76, 91.
Sepal, Definition of, 87, s 49.
Sequoia, 161, s 61, 99. See also "Big Tree" and Redwood.
Service Berry, s 113.
Sheep and Plants, 201, 203, s 22.
Shelf Fungus, 57, 59, Fig. 19, p. 58, s 35, 36.
Shepherd's Purse, 175, 202, s 68, 125.
Shield Fern, 81, Fig. 28, p. 78, s 46.
Shooting Star, 95, 97, 181, 198, Fig. 35, p. 96, s 54.
Shrubs, 108, 109, 177, s 64, 65.
Shrubby Mimulus, 138, 139.
Silene, 194, Fig. 70, p. 195, s 110, 119, 125. See also Indian Pink.
Silver-Back Fern, 77, s 46.
Silver Fir, s 100.

Sisyrinchium, 164, Fig. 61, p. 165 s 76, 123. See also Blue-Eyed Grass.
Skin, see Epidermis.
Skull-Cap, s 75, 86.
Skunk-Cabbage, s 104.
Skunkweed, 207, s 126.
"Sleeping" Positions, 64, s 38, 39.
"Smilax," (Myrsiphyllum), 164, s 101, 121.
Smut, 53, s 31.
Snapdragon (Mimulus luteus), 138, s 21.
Snapdragon (Antirrhinum), s 85.
Snow-Berry, s 65, 115.
Snow Plant, 51, 181, 198, s 120.
Soap-Root, 64, 66, 164, 183, 184, Fig. 22, p. 65, s 39, 116.
Solanum Douglasii, 123, Fig. 47, p. 122, s 21, 73, 114. See also Nightshade.
Solanaceæ, s 69, 73, 113.
Soldanella, s 120.
Solidago, s 25, 90.
Solitary Flowers, s 96.
Sonchus, s 96. See also Sow-Thistle.
Sorghum, s 106.
South America, 164, 179, s 115.
Sow-Thistle, 202, 210, 211, Fig. 73, p. 209, s 94.
Spadicifloræ, s 104.
Spanish Bayonet, see Yucca.
Spanish Moss, s 103.
Spanish Needle, 202, 210, s 92.
Species, s 87.
Speedwell, s 85.
Spergula, s 125.
Spermaphyte, s 42.
Spermatozoids, s 42, 45.
Sperm Cells, s 10, 50.
Spinach, s 110.

INDEX

Spirogyra, s 6.
Spore, 53, 57, **59**, **83**, 160, s **6**, **7**, 31, 32, **33**. **42**, **44**, **45**, 47, 48.
Spore-Case (Sporangia), 74-84, Fig. 30, p. 84, s 45-47.
Spore Distribution, s 45.
Spore Fruit, s 10, **32**, 33, 34.
Sporophyte, s 42.
Spurry, s 125.
Squash, 29, 31, **Fig. 7**, p. 30.
Stachys, s 86.
Squirrels and **Seeds**, 27.
Stamen, Definition of, 87, **s 49**.
Starch, 19, 38, **s 5**, **6**, 18, 19.
Star Thistle, 210, Fig. 72, p. 206, s 94, 125.
Star Tulip, s 74.
Stellaria, **s 55**, 125. See also Chickweed.
Stem Structure, s 58.
Stephanomeria, s 25.
Stipelia, s 114.
Stigma, Definition of, 87.
Stink-Horn Fungus, s 36.
Stipa, s 106.
Stipules (Pair of organs at base of leaf stem), s 59
Stomata, 36, 37, **42**, **43**, Fig 10, p. 56, **s 17**, **23**.
Storehouse, Underground, 39, 64-68, 72-85, 89, 93, 97, 110, 183, Fig. 22, p. 65.
Stramonium, **s 73**.
Strawberry, 179, **180**, **s 62**.
Strychnine, **s 114**.
Style, Definition of, 92.
Sugar Beet, **s 110**.
Sugar Cane, 168, **s 106**.
Sugar Pine, **s 100**.
Sumac, 177, **s 64**.
Sumatra, 60, **s 104**.
Summer Flowers, 183-199.

Summer Plants, 39-50, 196-199.
Sundew, 176, s 111, 112.
Sunflower, 44, 151-154, 201, **208**, Fig. 58, p. 152, s 91, 96.
Superior Ovary, 163.
Sweet Alyssum, 175, **s 68**.
Sweet Flag, **s 104**.
Sweet Potato, **s 114**.
Switch Plants, s 24.
Sword Fern, 81, Fig. 28, p. 78
Sycamore, 71, 104, 172, Fig. 38, p 103, s **28**, **41**, **57**, 59, 60.
Sympetalæ, s 113.
Symphoricarpus, s 65.

Talipot Palm, **s 104**, **105**.
Tan-Bark Oak, s 108.
Tansy, s 93.
Tapioca, 178.
Tar-Weed, 42, **44**, 197, 201, 207, s 25, 89, 92.
Taste of Plants, 41, **68**, **s 22**.
Tea Plant, 176, **s 112**.
Teasel, **s 114**.
Tellima, s 75.
Tendril, s 122.
Thistle, 151, **155**, 182, Fig. 59, p. 156, s 93.
Thistle-leaved **Salvia**, 146, 148.
Thrips, 92.
Thysanocarpus, s 68.
Thyme, s 114.
Tidy-Tips, 151, 154, **s 75**, 91, 125.
Tiger Lily, 163.
Toad-Stool, 57-59, Fig. 19, p. 58, s 34-35.
Tobacco, **s 73**, **113**.
Tocalote, **210**.
Tomato, **s 73**, **113**.
Topping Trees, s 57.
Tradescantia, s 104.
Transpiration, **s 14**, **23**.

INDEX

Transpiration Current, 37, 42, 43, 75-77, s 18.
Tree Mallow, s 65.
Tree Poppy, 112, 108, s 66, 67.
Trees, 70-71, 100-109, s 57-62.
Trees and Rain, 98.
Trees in Autumn, s 23, 27-28.
Tree Tobacco, 211, s 73, 127.
Trichostema, 207, Fig. 15, p. 49, Fig. 16, p. 50, s 26, 86. See also Blue Curls.
Trifolium, s 78, 79.
Tropæolum, s 81, 82.
Tropical Vegetation, 59, 60, 83, 84, 168, 170, 174, 176, 177, 179, s 45, 46, 109, 111, 112, 115.
Tropidocarpum, s 125.
Tubercles on Roots, s 30.
Tule, 162, 170.
Tulip, 164, s 101.
Tumble-Weed, 202, s 110, 126.
Tuna Cactus, 189-191, Fig. 67, p. 190, s 117.
Turkey-Weed, 46, 47, 177, 197, 201, 207, Fig. 14, p. 47, s 22, 25, 120. See also Trichostema.
Turnip, 175, 176, 200, s 68.

Ulva, s 8.
Umbelliferæ, 155-158, 178, 207, Fig. 22, p. 65, Fig. 60, p. 158, s 39, 75, 94, 95, 123.
Umbellularia, s 111. See also Bay.
Underground Stem, see Storehouse.
Upas, s 109.

Varnish on Leaves, s 26.
Venus' Fly-Trap, s 111.
Venus-Hair Fern, s 46.
Verbascum, s 85.
Verbena, 174, s 114.
Verbenaceæ, s 83, 114.

Veronica, s 85.
Vertical Leaves, 43, s 23.
Vetches, s 79.
Vicia, s 79.
Victoria Regia, s 111.
Violet (Viola), 93-95, 97, 175, 198, Fig. 34, p. 94, s 54, 111.
Virginia Creeper, 177.
Viscid (Sticky), s 26.

Wall-Barley, 204, 205.
Wall-Flower, Western, s 68.
Walnut, 31, 105, 172, Fig. 38, p. 103, s 13, 59, 60, 108.
Wandering Jew, s 104.
Washingtonia, s 104.
Wasps, 107, s 63, 85.
Water Cress, 175, 207, s 21, 68, 124.
Water Fern, s 47.
Water Fennel, s 6.
Water Hyacinth, 207.
Water Lily, 176, s 111.
Water-Net, 17-21, Fig. 1, p. 18, s 3.
Water, Uses to Plants, 39, 42, s 20. See also Transpiration Current.
Wax Palm, s 105.
Wax Plant, s 114.
Weaving Plants, s 41.
Weeds, 174, 199, 200-211, s 21, 27, 123-127.
West Indies, 174.
Wheat, 38, 162, 168, Fig. 9, p. 23, s 14, 106.
Whispering Bells, s 71.
White Forget-me-not, 121, Fig. 46, p. 120.
White Poppies, 112.
White Sage, 146-150.
Wild Cherry, Wild Grape, etc., see Cherry, Grape, etc.
Willow 40, 71, 100-104, 172, Fig. 37, p. 101, s 41, 59, 108.

INDEX

Wind Pollination, 102, 104, 105, 107, 108, 168, 169, 177, s 50, 60, 61, 126, 127.
Winter Plants, 61-71, s 37-41.
Woodwardia, 80, 81, Fig. 28, p. 78, s 46.
Woody Strands, 35, 37, 72, 162, 172, 177, s 17, 47, 58, 101.
Woolly-Back Fern, 80, s 46.
Woolly Breeches, s 72. See also Amsinckia.
Wormwood, s 25, 93.

Xanthium, s 91, 95.

Yarrow, s 92.
Yeast Plants, 52, s 30.

Yellow Forget-me-not, or Heliotrope, 121, 122, 201, 207, s 72, 124, 125. See also Forget-me-not and Amsinckia.
Yellow Mats, s 75, 95.
Yellow Pine, s 100.
Yerba Buena, s 86.
Yerba Mansa, 207, s 109, 124.
Yerba Santa, s 72.
Yucca, 162-166, Fig. 62, p. 167, s 24, 102-103. See also Frontispiece.

Zauschneria, Fig. 15, p. 49, s 26, 52. See also Fuchsia, Wild.
Zoöspore, s 7.
Zygnema, s 6.
Zygadenus, s 56.

ERRATA FOR SUPPLEMENT.

Page 8, line 17 ; for Neyreocystic, read **Nereocystis**.
Page 8, line 20 ; for Algæ, read **Alga**.
Page 8, last line ; for florescence, read **fluorescence**.
Page 21, line 4, for knot-grass. read **knotweed**.
Page 21, line 19 ; omit the word "first."
Page 21, fourth line from bottom ; for *nigraum*, read *nigrum*.
Page 21, third line from botton ; substitute comma for semi-colon after oak.
Page 24, line 4 ; insert comma after drought and omit it after parts.
Page 25, line 16 ; for *Malacothriz* read *Malacothrix*.
Page 25, line 27 ; for Helianthemum, read **Helianthus**.
Page 26, lines 10 and 16 ; for basis read **bases**.
Page 27, line 5 ; for prevail, read **prevails**.
Page 32, tenth line from bottom ; for the fields, read **these fields**.
Page 33, line 27 ; for spores, read **spore-fruits**.
Page 46, line 4 ; for Gynmnogramme, read **Gymnogramme**.
Page 58, line 12 ; for annular, read **annual**.
Page 66, line 7 ; for copo, read **copa**.
Page 66, fifth line from bottom; for Echscholtzias, read **Eschscholtzia**.
Page 67, line 15 ; for Eshscholtzia, read **Eschscholtzia**.
Page 70, eleventh line from bottom, for chapparal, read **chaparral**.
Page 76, third line from bottom ; for Fig. 6; also Fig. 2, read Fig. 61 ; also Fig. 22.
Page 80, line 23 ; for chapparal, read **chaparral**.
Page 81, last line, for Tropæolium, read **Tropæolum**.
Page 101, line 29 ; for Mersiphyllum read **Myrsiplyllum**.
Page 112, line 25 ; for Hemispheres, read **Hemisphere**.
Page 114, line 3 ; omit small.
Page 124, line 7 ; for *Elodia* read *Elodea*.

Borraginaceæ is twice written with one "r," a spelling authorized by the Century Dictionary, but the author had intended to use the more common spelling with two "r"s.

Foot-hill has several times crept in without its hyphen, and there are other errors and inconsistencies in the use of hyphens and commas that the author regrets, but does not consider necessary to enumerate.

www.ingramcontent.com/pod-product-compliance
Lightning Source LLC
Chambersburg PA
CBHW031423230426
43668CB00007B/418